# Integrated Science

## Volume 2

**Editor-in-Chief**
Nima Rezaei, Tehran University of Medical Sciences, Tehran, Iran

The **Integrated Science** Series aims to publish the most relevant and novel research in all areas of Formal Sciences, Physical and Chemical Sciences, Biological Sciences, Medical Sciences, and Social Sciences. We are especially focused on the research involving the integration of two of more academic fields offering an innovative view, which is one of the main focuses of Universal Scientific Education and Research Network (USERN), science without borders.

**Integrated Science** is committed to upholding the integrity of the scientific record and will follow the Committee on Publication Ethics (COPE) guidelines on how to deal with potential acts of misconduct and correcting the literature.

More information about this series at http://www.springer.com/series/16554

Hrvoj Vančik

# Philosophy of Chemistry

Springer

Hrvoj Vančik
Department of Chemistry
Faculty of Science
University of Zagreb
Zagreb, Croatia

ISSN 2662-9461 ISSN 2662-947X (electronic)
Integrated Science
ISBN 978-3-030-69223-0 ISBN 978-3-030-69224-7 (eBook)
https://doi.org/10.1007/978-3-030-69224-7



This Springer imprint is published by the registered company Springer Nature Switzerland AG
The registered company address is: Gewerbestrasse 11, 6330 Cham, Switzerland

# Preface

Serious and systematic investigations on the philosophy of chemistry became more intensive during the last 30 years, especially after the appearance of the two scientific journals dedicated to this field, *HYLE* and *Foundations of Chemistry*. Since I was involved in these investigations with some publications, I have initiated the new course *History and philosophy of chemistry* for undergraduate students at the University of Zagreb, Faculty of Science, Department of Chemistry. Experience in the education of students in such a new field was crystalized in the collection of themes which cover the main questions of philosophy of chemistry in the historical context. In this book, I offer my lectures, 13 of them, in somehow condensed form. First version of this book, which is the official university textbook of the University of Zagreb, has been published in Croatian in 2016 (Croatian title: *Dvanaest predavanja iz filozofije kemije*).

Philosophy of chemistry is considered not only as a branch of philosophy of science but also as a special branch of chemistry. As I will try to explain in the following chapters lectures, chemistry, as well as other sciences, are strongly context-dependent. This context comprises historical and cultural circumstances under which particular discoveries or concepts have appeared. Under the cultural circumstances, the most important are the philosophical thoughts that have fashioned science and art. But, as philosophy influences science, science reshapes also philosophy. One of the most important roles of the philosophy of chemistry is to investigate the philosophical and metaphysical concepts that emerge from chemistry. Gradual structuring of chemistry through history has opened and extended a series of philosophical questions, especially about the scientific methods, truth, identity, logics, analogy, etc. The question of the autonomy of science, the position of chemistry between physics and biology, is common to almost all the lectures. The autonomy of scientific disciplines and their hierarchical classifications are discussed within the theory of general complexity. The question about the demarcation between science and pseudo-science, as well as the problem of verification or rejection of scientific concepts and theories, is present through all the chapters.

In the first lecture-chapter, the scientific epistemology is discussed in relation to the classification of complexity levels and to the problem of holism and reductionism. The discussion is especially focused on the problem of the reduction of chemistry to physics. The hypothesis about the possible "anthropological" origins of physics and chemistry, as well as the first philosophical ideas that are the precursors of chemical concepts, is a scope of the second lecture. Among the central concepts in this consideration are the problems of element, principle, and change. Third and fourth lectures are dedicated to the first appearance of chemistry, the Alexandrian proto-chemistry, and alchemy. The practice of experimentation that has appeared in Hellenism is considered as an origin of chemistry, and as the predecessor of Arabian and European alchemy. Scientific revolution in Enlightenment, mechanical philosophy, new approach in scientific education, and internationalization-"objectification" of science is in the focus of the fifth lecture, with the intention to explain how the relationship between scientific practice and education is bidirectional: as science should be in the base of education, conversely, the educational methods provide special requirements to science to be organized in a specific way. The sixth lecture is focused on the importance of the choice of the investigating models, and how this model selection is critical for the formation of the new scientific paradigm. I demonstrate how the selection of combustion process as a model for the investigation of chemical changes led to the formation of the first chemical theory, the phlogiston theory. The seventh lecture is dedicated to the epistemological methods where the most important is the principle of *ceteris paribus*. All the scientific theories, laws, and concepts are valid only within the *a priori* given conditions and circumstances. Consequently, there are not natural laws but the laws of particular sciences. This idea led to the eighth lecture about the scientific revolution in chemistry in which the central figure is Lavoisier. In this chapter, the case study about the rejection of the phlogiston theory is used for critical discussion about the extent and limitations of radical falsificationism. The ninth lecture is rich with new ideas about general questions which can emerge from chemical reasoning; from the relationship between thinking in logics and thinking in analogy, till the problem of identity. The atomic theory is discussed as the consequence of the quantitative approach in chemistry, i.e. how it is possible to experimentize with invisible microcosm. The new definition of the notions of structure and element is in the focus of the tenth lecture. It is explained how the chemical epistemology based on analogies and classification can lead to the "invention" of the structure, and to its representation by chemical formulas and chemical nomenclature. Structural theory that origins from the theory of radicals and the theory of types is in the base of modern chemistry. I intend to show how the periodic law of classification of chemical elements is constructed by using the same methodology that was successful in the formation of the structural theory. Eleventh chapter is dedicated to some problems about the use of models in the discovery of the spatial configurations of molecules and chirality. The electronic structure of atoms and molecules, as well as the consequences of this discovery on the general ways of thinking in science, is the scope of the twelfth lecture. It is demonstrated how the knowledge about the limitations of the structural theory has appeared in the moment when the structural theory is transformed in the theory about reaction mechanisms. The last thirteenth lecture is a reflection of the first lecture, and it deals with the problem of chemistry in designing new substances, which have its scope in the appearances on the levels of complexity higher than chemical level. The problem of chemistry as basic science, and chemistry as auxiliary discipline is discussed.

Zagreb, Croatia

Hrvoj Vančik

# Contents

# About the Author

**Hrvoj Vančik** (Varaždin, Croatia, 1952) is a professor of chemistry at the Faculty of Science, University of Zagreb, Croatia. His scientific activity comprises the study of organic reaction mechanisms, especially thermally induced reactions in solid state. Most of his work belongs to the field of vibrational structures of carbocations and solid-state reactions of nitroso and polynitroso compounds. His work in the philosophy of science is focused on the theory of complexity in chemistry.

# Chemistry and Philosophy of Science 1

*At the centre of all things resides the Sun. Could we find a better place in this most beautiful of all temples, from whence this light illuminates all things at once? Rightly is it called the lamp, the spirit, the ruler of the universe. For Hermes Trismegistus it is the invisible god, Sophocles' Electra calls it the all-seeing. Thus, the Sun sits on its royal throne and guides its children, which circle it.*

Nicolaus Copernicus

**Abstract**

This chapter deals with the basic concepts of philosophy of science, and investigates the position of chemistry within this discipline starting from the time of Vienna Circle till the formal establishment of the philosophy of chemistry. The problem of the unity of science in relation to the appearance of particular scientific disciplines is discussed in the frame of the theory of complexity based on the idea that complexity is a hierarchy of structures from sub-atomic to biological. Besides the classification of sciences based on the fields of scientific interest, the alternative classification that follows from the complexity levels of matter is represented. Complexity levels are defined in the relation to the interaction energy between the particles of the lower level in the formation of the higher-level aggregates. The discussion about emergence, reductionism and holism is especially focused on the old problem of the reduction of chemistry to physics.

**Keywords**

Philosophy of science · Unity of science · Autonomy of scientific disciplines · Ontology · Emergence · Reductionism · Holism · Theory of complexity · Levels of complexity · Induction-deduction · Falsificationism

 
H. Vančik, *Philosophy of Chemistry*, Integrated Science 2,
https://doi.org/10.1007/978-3-030-69224-7_1

Ladies and Gentlemen,

The relationship between philosophy and science is, first of all, hierarchical. The intention of philosophy is building a general conceptualization of the structure and meaning of the universe, it constructs some basic patterns, which are reflected not only in scientific ideas, but also in human relations and in the society as a whole. However, the method of generalization of particular empiric regularities through generalized laws, what is the role of metaphysics, is in principle questionable. Namely, if some constructed concept or hypothesis, or even a language is more general, it becomes less precise. Following this idea, philosophical, say, metaphysical concepts cannot be immediately applied. Rather, they should be reduced to scientific laws and theories. The metaphysics could be evaluated just by using this criterion: good metaphysics ought to be reducible to physics! [1]

Conversely, the generalization of physics leads to metaphysics. This relationship physics-metaphysics is in focus of philosophy of science. In the broadest possible sense, we could consider the possibility of analogous relationship such as chemistry-metaphysics or metaphysics-chemistry. Other words, how chemistry could contribute to philosophy, which of chemical concepts could be generalized, or, which of metaphysical schemes have better examples in chemistry than in other sciences [2]. The question that emerges is following: would philosophy be the same as it is if chemistry would be absent from the spectrum of various sciences?

Philosophy of science plays also another even more important role: it is science about sciences, the metascience. Before discussing this aspect of philosophy of science, we have to tell something more about systematization (or classification) of particular sciences.

There are two independent modes of systematization of sciences. First system of classification is based on the fields of scientific interest: subject-oriented sciences, structural sciences, and metasciences [3]. Subject-oriented are sciences dealing with the perceptual experience of the nature and society, for instance physics, chemistry, biology, geology, anthropology, psychology or sociology. In focus of structural sciences are abstract ideas, structures, the ways of connections between entities, derivations, methods of conclusions. Their knowledges and results are fundamental basis for functioning of the subject-oriented sciences. Logics, mathematics, informatics, theory of automata, cybernetics, system theory, game theory, theory of formal languages, etc., belong to this group. Subject-oriented sciences have to use the methodologies of structural sciences for studying their subjects! Similar or analogous concepts that have its origin in structural sciences can be recognized even in very different subject-oriented sciences. This led us to an important mode of thinking, the use of analogies. As we will explain in following chapters, the thinking in analogies has been crucial in the historical development of chemical concepts [4, 5]. Within the third group of sciences we can find various metasciences, i.e. sciences about sciences such as philosophy of science, cognition theory, semiotics, sociology of science, etc.

Just to mention, there are also fields of sciences that have important public interest such as normative sciences (law, ethics, aesthetics), historical sciences (history, archaeology, bibliology), as well as special applied sciences, medicine, technology, etc.

Now, let me discuss an alternative mode of classification of sciences, which origins from the *degree of complexity* of natural objects. Since the term complexity is very difficult or almost impossible to define unequivocally, its meaning will be clarified successively in discussions later in this chapter. It could be argued that the complexity of nature, say physical nature, can be classified in a sort of *complexity levels* [6–8]. Let us imagine that the lowest complexity possesses elemental particles, the components of atoms. In that case the higher complexity levels comprise molecules, molecular aggregates, living cells, living beings following with eco-systems, and finally human society. If every of these levels could be investigated by means of the specific conceptual system characteristic only for this level, then such conceptual system forms authentic science (scientific discipline). For instance, biology is authentic science because it has been developed theoretical and methodological systems that belong only to biology, not to other sciences. Thus, chemistry could be structured on the same way, and it should be treated as an authentic science. Though, this question on the authenticity of chemistry remained open within the circle of philosophers of science for the long period of history [9]. To summarize, this hierarchy of sciences, if it is based on the levels of complexity, begins with physics, followed by chemistry, then biology, geology, anthropology, etc.

The long-lasting historical dilemma about the authenticity of chemistry as a scientific discipline [10] seems to be one of main reasons that philosophy of science did not involve chemistry, except in few particular case studies. Absence of chemistry in philosophy of science could also be explained by the historical domination of two traditional philosophical standpoints. By first opinion, chemistry is just a special case of physics, say, all the chemical concepts could be reduced to physical concepts, laws, and theories. Such position is called *reductionism*: the whole is nothing more but a sum of its components [9]. From the reductionistic point of view, characteristics of a molecule must be explainable if all the properties of the consisting atoms are known, or, the properties of living beings could be predictable by knowing their chemical composition.

The opposite concept by which the property of the whole is more than a sum of properties of the consisting components is called *holism*. For the instance, simple observations demonstrate that water molecule possesses particular properties that are not the properties of consisting atoms, hydrogen and oxygen. Consequently, properties of water cannot be predicted just by the knowledge about the nature of oxygen and hydrogen. These new unpredictable properties of water emerge only until the water molecule, i.e. the structure on the higher complexity level, is formed. These new properties are named *emergent* properties [11, 12]. Holistic behaviours on a particular complexity level characterized by emergent properties could be satisfactorily explained and described exclusively by conceptual systems, laws or theories that become a basis of corresponding authentic scientific discipline. Thus, the units of subject-oriented sciences with new characteristic emergent properties can be described and represented by using new emergent laws, which shape a conceptual system of particular authentic science.

In general, many of observed properties of substances are holistic. Let us take as example the contraction of volume of a mixture of liquids. The volume of the mixture obtained by mixing of 1 L of water and 1L of ethyl alcohol is less than two litters. The extent of such volume contraction is unpredictable from the knowledge about the properties of pure water and pure ethyl alcohol, it can only be found by measuring the volume after mixing. Thus, the volume contraction is an emergent property!

In summary, the interpretation of Universe formed through long history should be holistic. Thus, this interpretation is level-like, and it is divided in particular fields of knowledge such as scientific disciplines, or, from the other hand, artistic visions of the Universe and the surrounding world. However, we have to accept the fact, that the properties and structural organizations of the substance on the higher level of complexity are consequences of the properties of subunits on the lower level of complexity [11]. For instance, if the charge of electron and its fermionic property would be different than it is, this would reshape all the observed nature. From this point of view, the nature is reductionistic and explainable with causal bottom-up relationship: every variation on the lower level causes changes on the higher level of complexity. Or, the nature on the higher complexity level is causally related to the nature on the lower complexity level. Such reductionism is generally accepted, it is called *ontological reductionism* (οντος is being *per se*). Ontological reductionism is self-evident, so chemistry and complexity-higher sciences are in this sense reducible to physics. Why is chemistry nevertheless not a special case of physics? Such contradiction could be solved starting with the fact that emergent properties can only be described by special laws, which cannot be derived from physical laws. Analogously, biological laws of inheritance or reactions on environment cannot be reduced to chemical laws, for instance to Dalton's law of gravimetry or to the laws of organic reaction mechanisms. It would be very difficult even to imagine that the principles of organic synthesis are derivable from Schrödinger equation.

Intertwining of the holistic and the reductionistic aspect can be considered within the categories of *cause* and *effect* by taking into account the assumption that the appearances on the lower complexity level are causes of the appearances on the higher level of complexity, **but not *vice versa***![1] For example, the electronic structure of the molecule of the natural organic compound *indigo* is the cause that the substance indigo is coloured blue-violet (the emergent property). However, the same colour can also be property of other substances that consist from the

[1]#This statement will be relativized in the XIII chapter of this book.

molecules with completely different structure, as well as different electronic structure. Such cause-effect sequence is one-directional, i.e. the structure on the lower level of complexity always causes exactly defined appearance (say, colour) on the higher complexity level. On the other hand, the emergent property on the higher level (the blue-violet colour) is not dependent on the particular molecule, it can appear from quite different causes. Painters or other visual artists are interested primarily for the colour which they use on the artistic level of complexity, and not for the type of molecules or elemental particles that are behind this colour. The opposite causal sequence is not possible: blue-violet colour of a substance cannot be the cause of any electronic structure of the molecule!

Evolutional biologists are familiar with the similar problem that provoked the division on *Lamarckists* and *Darwinists*. While Lamarcksists maintained that the adaptations gained during the life of the individual living being cause the change and evolution of its genotype, Darwinists asserted that the evolution is possible only by causal bottom-up direction, i.e. from genotype to phenotype, and not vice versa. Genotype, the simpler complexity level, shapes the corresponding phenotype that belong to the higher complexity level.

Concepts, hypotheses, theories, and laws as the basis of the scientific methodology represent the *epistemology* of particular scientific discipline (επιστημη means knowledge) [1]. Every field of science consists from *ontology* component, i.e. the essence or the being of their subject, and the *epistemology* component, which serves to outline representations and explanations of the nature of the subjects. Authentic science, like chemistry or biology, is holistic not because their phenomena are independent on the phenomena of more elemental science, but because it produces autonomous *laws and theories* that are not reducible on the laws and theories of the discipline that deals with the less complex subjects. In summary, if we talk about the holism-reductionism dilemma in sciences, the reductionism is valid only in the ontology sense. Holistic structure emerges from the epistemology component of science, which is a core of the authenticity of scientific discipline.

Let me now focus on the question about the rational, i.e., possible physical interpretation of the previously mentioned classification based on the complexity. Arranging of complexity of the nature in levels [6] is evidently a sort of general lawfulness that could be explained within the particular structural science, for instance, within the theory of systems. In the search of the above-mentioned lawfulness it would be reasonable to ask the questions such as: is there a clearly defined physical parameter characteristic for the corresponding complexity level, or, is it possible that complexity can grow *ad infinitum*, or, how sharp is the borderline between complexity levels? Complexity levels as hierarchy of nature were proposed by many authors by using various criteria. Emmeche et al. have proposed four levels: physical, biological, psychological, and sociological [12]. However, the unambiguous criterion for the arrangement of complexity in levels must be based on the particular physical parameter, which has been generally used in the explanation of the most of the structures in the Nature. This parameter is energy. Suppose that two entities on the lower complexity level unify in the new structure that belongs to the higher complexity level. For instance, two atoms are associated forming the corresponding molecule. The consequence of the formation of molecule is the decrease of total energy of the system because the two-atomic molecule always possesses less energy than the sum of energies of two isolated atoms. This energy difference is called the energy of chemical bond. Conversely, if we wish to split the molecule in the isolated atoms, we have to add to it the energy equal to the chemical bond energy. On the sub-atomic level, association of proton and neutron in corresponding atomic nucleus follows the same rule with the release of large amount of energy (nuclear energy). If the atom would be exposed to such amount of energy, it would decompose in protons and neutrons. In principle, the chemical bond energy and nuclear energy, both could be named by using *terminus technicus* energy of interaction.

Beginning from the lowest complexity level, for instance from elemental particles, then following with atomic level, then molecules and molecular aggregates, etc., each transition between different levels requires corresponding energies of interaction between entities. Interaction energies required for the transition to the higher complexity level are represented in the following figure.

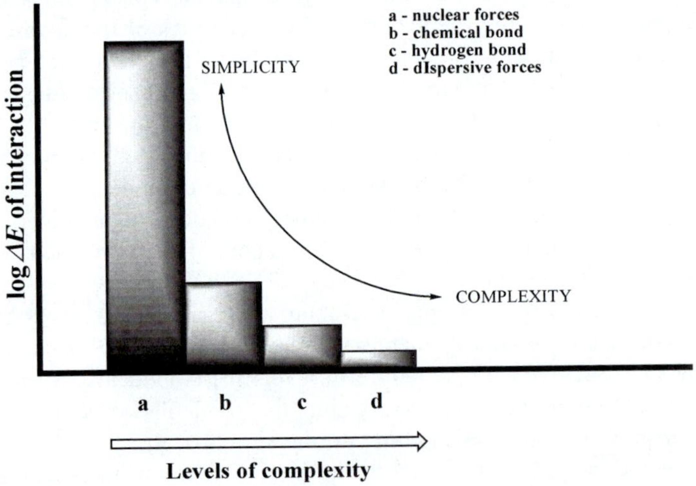


**Complexity classified in the levels in relation with the energy of interaction of particular categories of particles: atomic nuclei, molecules, and molecular aggregates**

As it is shown on the upper figure, the interaction energy for the formation of atoms from elemental particles is nearly one *million-fold larger* than the energy of chemical bond. On the other hand, the energy of aggregation of molecules by hydrogen bond is only *five-fold lower* than the bond energy, and the energy of arrangement of molecules in liquid or in molecular crystal is still much lower and it could be *almost negligible*. If the relationship between energies of interactions and corresponding complexity levels is represented on such a quasi-quantitative way, then one of possible conclusions could be: complexity is asymptotically approaching to some limit. Does it mean that complexity cannot grow *ad infinitum*? But now, on the left side of the diagram where the "simplicity" dominates, the interaction energy grows asymptotically to the infinity. Does it mean that the absolute simplicity is also impossible because it would require the infinite amount of energy? Accordingly, absolute simplicity and absolute complexity are not possible. The level of complexity covered by chemistry is just in the middle of this diagram—chemistry is the central science!

The concept of the complexity levels that I have represented based on the energy of interaction between units is not the only possible. This energy criterion deals only with the stable particles vibrating around the potential energy minimum, i.e. atomic nuclei, atoms, molecules, etc. However, in chemistry, the most important are not so much the structures but rather changes of these structures. But the changes of structures, the chemical transformations, depend not on the stable structures but rather on the structures of the transition states that appear not in the potential energy minimum. Transition states are characterized with the structures appearing in the high-energy saddle-points where their vibrations are not explained with the rational numbers but with the imaginary frequencies. Since the direct classification of the transition states is almost impossible, as the criterion for systematization we can use the experimental observable that depends on the free energy of these transition states, i.e. the rate of chemical reaction.

Good models for studying the "shift" from chemical to biological systems are chemical reaction called hydrolyses. In contrast to the reactions in non-living nature, the reactions in living beings are catalysed by enzymes. Now, let us compare the rates of hydrolyses occurring without catalysts with those catalysed by enzymes. As it is shown in the following figure [13]. while the noncatalyzed reaction rates (rate constants) are spanned within the **wide range** from $10^{-1}$–$10^{-15}$ $s^{-1}$, the catalysed reactions appear within the **narrow range** of rates between $10^{5}$–$10^{8}$ $s^{-1}$. The reaction rates of nuclear processes, studied by quantum mechanics in the physical standard model, belong to much higher range of values than in the case of uncalalyzed chemical reactions. In this way, we have three levels of complexity. However, there is another more interesting point. The emergence of life is characterized by **narrowing the range of the rates of chemical reactions!**

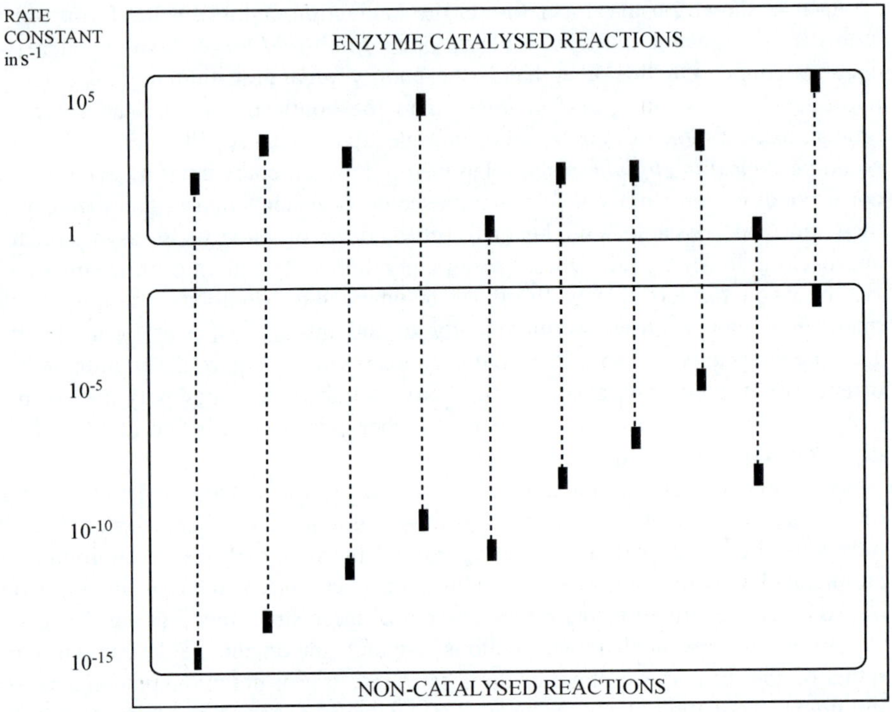


**Comparison of the enzyme catalysed and non-catalysed reaction rates. The figure is construdted from the data from ref [13]**

In the philosophy of science, we can simply conclude that the research within every complexity level requires special epistemology that is not reducible to the epistemologies from other levels. This could be convenient explanation why science consists from special disciplines such as physics, chemistry, biology, etc.

Consideration of the authenticity of chemistry starting with complexity of the nature and emergence of irreducible laws and concepts, as well as absence of chemistry in classical philosophy of science, must involve also questions about the historical circumstances within which appeared philosophy of science as an original discipline. Circumventing the detailed analysis of the history of philosophy of science, because it is a matter for an extra book, we must focus on those facts that could be most important for studying the relationship between chemistry and this philosophical discipline.

As I will demonstrate later in the lectures, it seems that chemistry through most of its history appears close to the borderline between science and "non-science", or, between science and "skill". From the 13th century till the scientific revolution after the Enlightenment period, chemistry has even not been a part of the university education. Reasons for such position of chemistry cannot be associated only to the structure of society, but also to the statements of philosophy of science. Clear indicator of the disrespect of chemistry we can find already in statements of some of the greatest philosophers. **Immanuel Kant** (1724–1804) divided science, in principle, in two categories: exact sciences and skills, say half-sciences [2]. By Kant, exact sciences represent their concepts, laws, and theories by using mathematics, because it is the mathematics that interprets the Kantian categories of mind: time and space. All the other disciplines belong to a sort of mixture of skills and pure empirical experiences. [14].Since in the time of Kant, the 18th century, chemistry was almost entirely qualitative, it has evidently been categorized as a half-science or just as a skill. However, in its later work, *Danziger Physik* (1785), Kant had admitted that some of particular philosophical concepts could be properly demonstrated by using analogy with chemical empirical laws [15]. Later, Kant has also added the name of chemist Lavoisier on the same list with the names of Archimedes and Newton! It should also be mentioned the personal contribution of Kant directly to chemistry. **Jeremias Benjamin Richter** (1762–1807), defended his dissertation about the application of mathematics in chemistry under the Kant's supervising at the University of Königsberg. Richter's work was very influential on the development of modern chemistry. This will be discussed in later lectures. As I have already mentioned, Kant's affinity to exact sciences based on mathematical representations origins from his philosophical system of categories of mind [16]. Space and time as well as the principle of causality are innate categories. Since they exist *a priori* in human mind, our cognition of the external world is possible. Today, we would perhaps say that there could exist special genes bearers of the time-space categories. Even some recent theories of cognition have revitalized, although on different way, such Kantian approach. From the modern point of view, the main disadvantage of Kant's theory is not so much in the idea about the innate categories of mind but rather in his concept of time and space used strongly within the frame of the Newtonian and Cartesian physics and mathematical interpretation. Such concepts of time and space are today very incomplete after the development of relativistic theories and quantum mechanics.

About the possibility of mathematical representation of the perception of time and space, in contrast to impossibility of mathematical representation of perception of other sense data such as colour, heat or taste, I will discuss in lectures that follow. Let us suppose that just the *convenience of different senses to be interpreted mathematically* could be the main historical origin of the development of the ancient proto-science in two directions, leading to physics and chemistry.

One of the classical philosophers who the best models for his metaphysical concept has found in chemistry, was **Georg Wilhelm Friedrich Hegel** (1770–1831). Hegel has used the term "chemic" for the properties of entities that are not absolute, but rather the properties which emerge only as a consequence of their interaction with the environment [17].Thus, all the observable characteristics of subjects are emergent! In addition, Hegel did not reject the idea about the innate categories. Rather, he did not accept that these innate categories are time and space.

Instead, he introduced the categories of dialectics, which are more related with the *changes* than with the *structures*. August Kekulé, one of the greatest chemists in history argues:

> **die Chemie ist also die Lehre von den stofflichen Metamorphosen der Materie. Ihr wesentlicher Gegenstand ist nicht die existierende Substanz, sondern vielmehr ihre Vergangenheit und ihre Zukunft ... [Chemistry is thus the theory of the material metamorphosis of matter. Its essential object is not the existing substance, but rather its past and future ...] (Kekulé 1861–1882, 1, p. 3).**

The historical beginning of the systematic research in philosophy of science is difficult to determine. Tentatively, one of pioneers in this field was **Ernest Mach** (1838–1916), the professor of mathematics at the University of Graz and physics at the University of Prague. After invitation from the University of Vienna, Ernest Mach has established new cathedra for *inductive sciences*. As I will argue latter, the method of *induction* as a way for the construction of scientific concepts and theories was in focus of the first organized group of philosophers in Vienna, which reached its highest activity during the first half of twentieth century. This plead of philosophers of science and theoreticians of knowledge is known as the *Vienna Circle*. Let me mention just some of most important names of the Vienna Circle: **Moritz Schlick**, **Hans Reichenbach, Rudolf Carnap, Ludwig Witgenstein**. Later I will talk a little bit more about **Karl Popper** (1902–1994), who actually declined from the ideas of Vienna Circle by rejecting induction as a method for constructing new scientific knowledge.

The basic idea that has been established within the Vienna Circle was that the empirical experience is the main starting point from which the scientific concepts can be created. This pathway from the experience to the conclusion or formation of hypothesis is called *induction*. Scientists and philosophers, who follow such a viewpoint belong to *empiricists*, or *positivists*. This positivistic approach, which is basically in contrast to Kant's opinion, affords that human being does not possess *a'priori* structures in its mind. Rather, all what the human absorbs from the nature in its mind is achieved by the experience, observation: by nativity, human mind is merely like an empty blackboard, «*tabula rasa*», in the words of **John Locke** (1632–1704) who has been one of the establishers of empiricism. The synonym *positivism* origins from **August Comte** (1798–1857), who has proposed the idea that the history of science could be divided in three periods, *mythopoetic*, *metaphysical*, and *positivistic*. The last period is for Comte the ultimate. Since the philosophers of the Vienna circle accept the fact that the theory in its final form could be very far from the immediate experience, they require that the experience should be transformed to the theory by using logics as a mediator. Scientific cognition must be derived from the empirical facts by a series of logical operations. Accordingly, the Vienna circle philosophers are also known as *logical positivists*. The most radical use of logical positivism is known from the work of **Ludwig Wittgenstein** (1889–1951), who even negated the reality of any philosophical problems: all the research in philosophy should be focused on the analysis of the logics of language, while all other considerations are pure play of words. **Bertrand Russell** (1872–1970) has transformed the "play of words" in the mathematical logics. Russell and his Ph.D. supervisor **Afred North Whitehead** (1861–1947) have finally established mathematical logics in their masterwork *Principia Mathematica* published for the first time in the year 1910.

The most prominent opponent to Vienna empiricism was Karl Popper, who has argued that the scientific cognition can only be achieved by *deduction*, rather than by induction [1]. The frameworks of a new proposed theory does not generally emerge from the experience, but rather from quite different causes, for instance from experiences from other scientific fields, personal experiences, social environment, intuition etc., and this theory must have such a form that it could be *falsified by deduction*. Accordingly, the basic characteristic of a good theory is not in its property to be confirmed, but rather in its property to be falsified! Bad theories, pseudoscientific ideas and dullness are irrefutable! For Popper, falsifiability should be self-contained in the theory on the way that it enables corresponding verification test that should be satisfied. If the theory does not pass the test, it should be rejected as a false. For instance, physical law that in *vacuo* all the objects fall down by the same velocity is good law because it could be *tested* experimentally. The advantage of this approach is that the verification of the theory is much more exact because it is always possible to find an infinite number of proofs, but only one experiment is enough to reject the theory.

To be historically correct, we have to mention that this idea is not originally invented by Popper. It has been shaped fifty years earlier by **Pierre Duhem** (1861–1916) who has argued that the scientific theory is an abstract construction that should be verified by observation and by the experiment. Falsifiability method proposed by Duhem is more appropriate in the scientific method since it is not so radical as the method of Popper. In the following chapters, we will discuss the examples where the Popperian radical falsificationistic approach could be counterproductive. Further extension of the Popperian method, which has been proposed by **Imre Lakatos** (1922–1974) will also be discussed later in this book.

In the history of science there are plenty of examples in favour of the Duhem-Popper falsificationistic method. For instance, **Nicolaus Copernicus** (1473–1543) did not invent the heliocentric planetary system by induction (i.e. by experience) from the observations of movements of heaven bodies, because the astronomy at the time has not an accessible observation data in favour of the idea about the rotation of Earth around the Sun, rather than Sun around Earth. The reason that moved Copernicus to heliocentric system was in his philosophical conviction rather than any kind of precise observation or measurement! Copernicus was part of the philosophical circle of *Neoplatonists*, the doctrine which initiated the philosophy of Renaissance. This version of Platonism appeared in the second and third century by ideas of **Ammonius Saccas** (3rd century B.C.), and it has been finally developed by the roman philosopher **Plotinus** (nearly 204–270 B.C.). In the frame of the Neoplatonist point of view, all the Universe emerges, it *emanates*, from one abstract, and infinitely distant point. In the Renaissance paintings, this point appears as a centre of perspective. By thinking in analogies, such point in astronomy is the Sun, and Copernicus as a Neoplatonist has been convinced that Sun should be in the centre of Universe! Epistemic deduction by which the heliocentric system could be finally tested has used the time of few centuries, and the significant development of astronomical methods. Though, as we will see later in the case studies from chemistry, radical Popperian falsificationism in evaluation of scientific concepts could also slow down the development of science.

The deeper insight into the debates within the Vienna Circle reveals that almost none of the case studies about the structure of scientific theories has been used from chemistry. Does it mean that for the philosophers within the Vienna Circle chemistry is not a science? Some of reasons for the absence of chemistry were undoubtedly also in the sort of education of the most important scientists in this Community. There we can find mathematicians, physicists, biologists, but no chemists! Since the Vienna Circle was the main melting pot from which philosophy of science has emerged, the nonexistence of chemistry could be assumed. Radical reductionistic epistemology in the Vienna Circle includes also the statement that science is unique, hence, there are no distinct disciplines, since all the knowledge could be reduced to physics [18]. This has been the central idea of a series of scientific conferences under the title *International Congress for the Unity of Science*, which have been held between the years 1931 and 1940. Discussions on these meetings were focused on logics, mathematics, physics, biology, psychology, even economy and linguistics, but there was no chemistry (neither geology).

Among the first scientists, who have used chemistry for study philosophy of science were **Friedrich Wilhelm Ostwald** (1853–1932) and already mentioned Pierre Duhem. Both were known as scientists and philosophers who have been focused on the epistemology that emerges from the research in classical thermodynamics rather than from classical mechanics. For Ostwald and Duhem the background of all the Universe is energy, and not the mater mechanically structured from atoms. In contrast to Duhem, Ostwald has changed his opinion regarding the existence of atoms only after the year 1897, when Thomson discovered electron (**Sir Joseph J. Thomson,** 1856–1940). In spite to this, Pierre Duhem was one of the first scientists who studied systematically chemical epistemology by analysing the appearence of the structural theory in the historical context. For Duhem, *science is not understandable without the systematic study of the history of science*. Consequently, philosophy of chemistry can be developed only within the historical reconstruction of scientific discoveries.

Entering of chemistry in the philosophy of science through the ample door appeared after the work of Thomas Kuhn (**Thomas Kuhn**, 1922–1996) who was one of the establishers of *sociology science*. In his masterwork *The Structure of Scientific Revolution* [19]. Kuhn investigate the structure, development, and functioning of science by using chemistry as a case study. Kuhn has accepted the Duhemian idea that the development of science is continuous, but he also asserts that this development is oscillating between the periods which he calls *normal science*, and the periods of *scientific revolutions*. During the period of the normal science, the scientific activity is focused on the problem solving within the bundle of stable scientific laws, concepts, and theories, which Kuhn calls a *paradigm*. For instance, before the Einstein relativistic theory, the practice of science took a place within the paradigm of the classical mechanics, i.e. within the pattern of the Newton theory. Scientific revolution has appeared as an extension of the Cartesian space and time to the relativistic theory that became the basis of the new paradigm. After that, again, a history of science has continued through a period of the normal science. Thus, by Thomas Kuhn, abandoning of the old and growth of the new paradigm is the scientific revolution. Interestingly, the case study for the explanation of scientific revolution that Kuhn has used is the change of paradigm by which **Antoine Laurent de Lavoisier** (1743–1794) has established basic concepts of the modern chemistry. Such a way, in the second half of the past century, chemistry finally comes as a part of the research in philosophy of science. In addition, in his book *Science in Action* (1987), **Bruno Latour** has represented a list of the most important scientists on which we can find Crick, Mendeleev, Pasteur, Einstein, Newton and Copernic. Half of those could be considered as chemists. At the end of Century, philosophy of chemistry has been finally organized as an egalitarian branch of philosophy of science [20, 21]. On the Conference in Ilkley (United Kingdom), which has been hold in 1997, the **International Society for Philosophy of Chemistry**, the organization that gather scientists and philosophers dealing with chemistry was established. In this period, two scientific journals focused on the philosophy of chemistry have appeared, **HYLE** (1995) edited by Universität Karlsruhe, and **Foundations of Chemistry** (1999) , which is today edited by Springer-Verlag.

Kantian concept about the proper and improper sciences, which was based on the degree on which it can use mathematics, undoubtedly has roots in the old history of systematic activity in the scientific knowledge. Independently whether we accept empiricism together with its inductive approach, or we accept Duhem-Popperian deductivism, the cognition that on the social level is called science origins from the senses as mediators between the mind and the outer world. Interestingly, senses as well as scientific disciplines can also be divided in two groups as regards the procedure by which they could be analysed. Senses of time and space belong to the first group, similarly to the Kantian categories, and they naturally can be associated with the quantitative way of thinking. These senses are the grounds of sciences which use mathematics as the main methodology: astronomy, geometry, arithmetic, mechanics, i.e. all that we today attribute to physics. Second group of sense data such as heat, colour, smell, or flavour cannot be used for the objective analysis because their nature is qualitative and there has been no such mathematics that could be used for their rationalization. There is also a fundamental difference between the approaches in cognition which follows from these two groups of senses. While the space and time could be measured, and from these measurements it is possible to construct abstractions on over ever higher levels until the formation of generalized theory, it has been not possible, at the least in the time of protoscience, to make experiments with the phenomena such as the events observed on the sky. On the other hand, colour, taste or smell were phenomena not measurable at the time, but convenient for experimenting, for instance melting of ores, without the possibility to construct any general theory. Here we offer suggestion that these two groups of senses represent the background for the parallel formation of the two scientific disciplines, physics and chemistry.

From the very beginning, physics (or better to say proto-physics) has been connected to astronomical observations, which were interpreted by mathematical representation with unequivocal formulas. Since with the starry sky is not possible to do experiments till the appearance of modern science, experiment was not the very traditional method in physics. In contrast, chemistry (or better to tell proto-chemistry) has been based exclusively on experiment, the mastery of simulations of natural processes. Preparation of metal in the metallurgical oven could be regarded as an emulation of the process which normally proceeds in the interior of Earth [22]. In this way, the tradition of chemistry was initiated with the immediate manipulation with the substance! Systematization of experimental perceptive experience guided to the formation of a sort of abstraction that is not mathematical, it is rather non-rational, and, as we will see later in the discussion about alchemy, it is often open to quite another goal: the immersion of the experimenter in the deeper *esoteric* levels of its own consciousness. However, these experiments, as a result of the mastery of manipulation with substance, have contributed also to the real practical world, they had also an *operative* or *exoteric* character. This ambiguity between the esoteric and exoteric character or the proto-chemistry and alchemy has for the long period in the history been one of the main reasons that chemistry was not regarded as a serious science.

As I wish to demonstrate, chemistry and physics have different heritages for which we could speculate that originated from the two groups of senses, which were by their nature diverse and complementary. Yin and Yang, the mathematical *ratio* and the experiment were two intellectual approaches that have constructed the modern interpretation of the Universe. While physics creates abstractions of the hardly imaginable forms of time and space, chemistry penetrates in the abstract profundity of comprehension of the ontology of substance. Chemistry emerges on the borderline between microcosm and macrocosm.

From the historical point of view, chemistry could even be regarded as an authentic science because it has its own epistemic origin. One of the concerns regarding the authenticity of chemistry is focused on the question of its ability to formulate *laws*, at the least in the way as it is possible in physics [23]. The answer on this objection could be possible only after we ascertain which is the nature of laws in science. Is the possibility of mathematical formulation the only way to formulate the scientific laws, or the laws could have also quite different modes of description? First of all, it should be mentioned that some of concepts and laws, which are today regarded as physical laws, originally appeared in chemistry. For instance, the credit to the formulation of the *law of conservation of mass* should be given to Lavoisier, because chemical experiments were in his time already enough quantitative to serve as a convenient model for the immediate verification of this principle. Quantitative investigation of the transformation of the precursor into the product substance has been a relevant approach for testing this law.

*Phlogiston* theory, [24] (which will be discussed in later lectures in this book), that has appeared at the beginning of the 18th century describes the combustion of some stuff as a release of the hypothetic substance called *phlogiston*. For instance, by combustion of a metal, metal loses phlogiston and transforms in the lime (the historical term for oxide). To recover the original metal, the lime should be heated with substances that are reach of phlogiston, like coal. Such a way, phlogiston is returned back to the metal. Thus, phlogiston is a *chimeric stuff* that could be exchanged between various substances. In words of modern chemistry, the release of phlogiston is oxidation, and addition of phlogiston is a reduction. If we are thinking in analogies, we could easily replace the word phlogiston with the word electron, and get the concept of oxidation and reduction. The role of phlogiston for chemists in 18th century could be compared with the role of electron after its discovery two centuries later.

The triumph of physical science in the history was certainly the appearance of quantum mechanics, thanks to Heisenberg 1925, and Schrödinger 1926. Since the principles of quantum mechanics will be discussed in the lecture about chemical bonds and reaction mechanisms, here I present only the basic principles. In the frame of quantum mechanics, the system is outlined by one infinitesimally complex and abstract mathematical function that represent a «superposition» of all the possible configurations. Let us imagine a system in a closed room. Since we don't have the insight in the events in this space, it could be concluded that in this room is a superposition of all possible scenarios. By opening the door of this room, we can find the configuration of existing states. However, *by this action*, opening the door, the superposition of all the possible configurations is destroyed, and only one state is detected! In other words, opening of the door caused a «collapse of superpositions» and the emergence of only one configuration. In mathematical interpretation of quantum mechanics, the *function* that describes the set of all the states has been subjected by *operator*, i.e. opening of the door, and we got the concrete visible state, but in the same time, we have cancelled the superposition. Thus, the wave function for electrons describes all the possible states of electrons, but by *decision* to use experiment for measuring their energy, say by action of the operator of energy, the superposition is cancelled and we provide a spectrum of measurable energies, so called *eigenstates*. Reception of the world by such method *depends*, accordingly, *on our selection* of the experiment, i.e. about the use of the operator that provide information of interest. Since all the other states of the system collapsed by action of operator, it is not possible to act by another operator on the same system. It follows that experimental measuring of one property (for instance energy) has the consequence that all other properties (for instance the position of electron in the space) become uncertain and undeterminable. Such approach is in the core of the quantum mechanics, and it is called the *Heisenberg uncertainty principle*.

The question is, whether chemistry, independently on the developments of physic, has formulated similar concept, at the least on the metaphysical level? During the first fifth of the last century, chemists were preoccupied by interpretations of molecular structures, and by the principles of the changes of molecular structures in chemical reaction. One of the most contradictory and unsolved problems of classical structural chemistry has been the problem of the structure of the benzene molecule. Robinson and Arndt, the scientists about which I will talk in one of the last lectures in this book, have proposed that the structure of the benzene molecule (and later the structures of a series of similar molecules) is better described as a "superposition" of higher number of classical structural formulas (see the Figure below). However, for the description of a particular reaction way that is under the investigation, we are constrained to use only the particular of these structures, say it could be a sort of the "collapse of such superposition".

Clearly, scientific revolutions from quite unknown reasons appear simultaneously in different fields of research and intellectual activities.

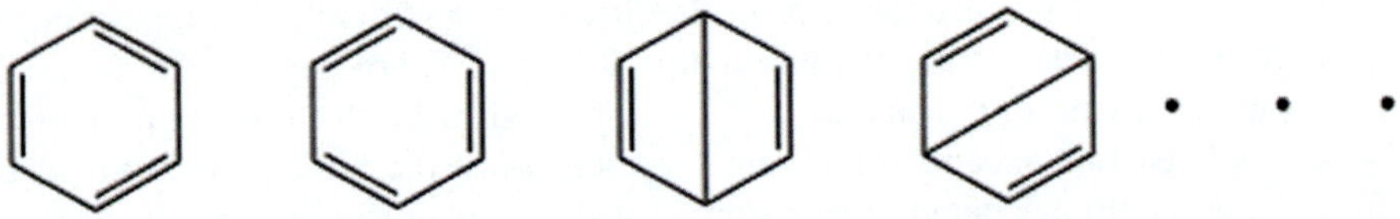

**Representations of benzene molecule proposed by Robinson and Arndt. Every of the structures makes sense only in combination with other structures**

By looking on quantum mechanics and mechanistic chemistry from the metaphysical level, the same concepts of reasoning appear, in spite to the historical fact that they were developed by quite different epistemologies, from the mathematical manipulations, or from the abstractions of molecular structures that have developed in chemistry simply by systematization of substances.

As I have already discussed, the concerns about chemistry as an authentic science were based also on the question of the ability of this discipline to formulate laws. Do chemical laws exist? First of all, it should be necessary to explain the nature of scientific laws. Are the scientific laws exclusively mathematical formulations? Isn't chemistry the discipline, which has extended the definition of scientific laws from the strict mathematical formulations to other modes of abstractions, such as the creation of special languages? For instance, is it possible to use every particular chemical reaction mechanism as a sort of chemical, thus, scientific law?

To conclude, physics and chemistry origin from different groups of senses of the surrounding, the quantifiable senses of space and time, and the senses that cannot be quantified simply, colour, taste, heat etc. This is the base from which the contemporary science has been developed.

## 1.1 Conclusions

The adequacy of the idea about the autonomy of chemistry as science follows from the analysis of the complexity levels of matter. Such a hierarchical structure of complexity is constructed starting from the energies of interaction between the entities (particles) of lower level to form particles on the higher level. Autonomy of science, as well as autonomy of chemistry is justified by the fact that every scientific discipline develops its epistemology starting from the emergent properties and processes occurring within its complexity level. Although chemistry was almost absent in the early development of philosophy of science because the main philosophical circle in history was oriented to other sciences such as physics or biology, and because of the prevailing idea about the proper sciences based on strong mathematical representations, its role in the shaping of the deepest concepts of modern science is out of questions. Chemistry has its own epistemology and the system of models for the representation of nature: system of elements, valence, structure, as well as the interpretation and practice of transformation of substance.

## References


1. Popper K (1963) Conjectures and refutations, London
2. van Brakel J (2000) Philosophy of chemistry. Leuven University Press, Leuven
3. Vollmer G (1990) Evolutionäre Erkenntnistheorie. S. Hirzel Verlag, Stuttgart
4. Duhem P (1991) The aim and structure of physical theory. Princeton University Press, Princeton, NJ
5. Needham P (2000) Atomic notation and atomistic hypotheses. Found Chem 2:127–180
6. Vančik H (2003) Philosophy of chemistry and limits of complexity. Found Chem 5:237–247
7. Vančik H (2004) Philosophy of chemistry and generalization of the concept of complexity, in chemistry in the philosophycal melting pot. Peter Lang, New York
8. Simon HA (1996) The sciences of the artificial. MIT Press, Cambridge, MA
9. Primas H (1983) Chemistry, quantum mechanics and reductionism. Springer, Berlin
10. Vančik H (1999) opus Magnum: An outline for the philosophy of chemistry. Found Chem 1:241
11. Scerri ER (2007) Reduction and emergence in chemistry: two recent approaches. Philos Sci 74:920–931. See also: Scerri ER (2007). The ambiguity of reduction. HYLE 13:67–81
12. Emmeche C, Koppe S, Stjernfelt F (1997) J Gen Phyl Sci 28:83–119
13. Wolfenden R (2006) Chem Rev 106:3379–3396
14. van Brakel J (2005) Kant's legacy for the philosophy of chemistry. In: Scerri E, Baird D, McIntire L (eds) Boston studies in the philosophy of science. Springer, Berlin
15. Blomme H (2015) Kants conception of chemistry in the Danziger Physik. In: De Gruyter RC (ed) 2015 Reading Kants Lectures
16. Tetens H (2006) Kants "Kritik der reinen Vernunft": ein systematischer Kommentar. Reclam, Stuttgart
17. Ruschig U (2001) HYKE 7:5–22
18. Hettema H (2012) The unity of chemistry and physics: absolute reaction rate theory. HYLE 18:145
19. Kuhn ThS (1970) The structure of scientific revolutions. University of Chicago Press, Chicago

20. Schummer J (2006) In: Baird D, Simon HA (1996) The sciences of the artificial. MIT Press, Cambridge, MA
21. Scerri ER, McIntire L (2015) Philosophy of chemistry, growth of a new discipline. Springer, Dordrecht
22. Eliade M (1977) Forgerons et alchimistes. Flammarion, Paris
23. Feynman R (1965) The character of physical law. Cox and Wyman Ltd., London
24. Stahl GE (1715) Opusculum Chymico-Physico-Medicum. Halle et Magdeburgiae, 1740

# 2 Perceptions, Experiments and Principles

*Man is the measure of all things of things that are, that they are, of things are not, that they are not.*

Protagoras

**Abstract**

The hypothesis about the possible "anthropological" origins of physics and chemistry is represented. It is argued how chemistry and physics have their proto-origins in different kinds of the sense data: time and space on the one side, and taste, colour, heat, smell, etc. on the other side. Physical way of thinking was connected with senses of time and space, which can be measured, and represented mathematically. The nature of other, more "soft" sense data is such that measuring and any mathematical representation was in ancient science impossible. Chemistry has appeared as an experimental skill of manipulation with these qualities. This leads to the earliest philosophical ideas that are the precursors of modern chemical concepts: element, principle, and change.

**Keywords**

Sciences and senses · Sense data · Origins of physics and chemistry · Elements · Principles · Elementary stuff · System of four principles · Transmutation of elements · Experiment-theory

Inspired by a questionable notion about the proper and non-proper sciences, we have proposed the idea that the proto-origins of physics and chemistry could be in the division of perceptions in those that can be analysed mathematically, and in those, for which mathematical interpretation was not possible in the early history of



H. Vančik, *Philosophy of Chemistry*, Integrated Science 2,
https://doi.org/10.1007/978-3-030-69224-7_2

science, but which seemed convenient for experimentation. The exact time in the history when experiments have emerged from practical crafts, or in other words, the time when manipulation with the substances has changed its role from practical to metaphysical is difficult to determine. As new speculative systems always appear by amalgamation of various views, the philosophical activity that has been transformed to chemistry appeared by intertwining of everyday craftsmen skills with the already developed philosophical systems.

How to rationalize the speculations about the nature of the substance? How to interpret the immeasurable and transformable stuff, and how from this interpretation to shape a particular world-view? What are categories of matter? Appearance of the metaphysics of substance is known in almost all of the ancient civilizations, but we will focus our discussion on the ideas, which were appeared on the Eurasian Mediteranesn region. Metaphysics, which emerged from the immediate manipulation with substance, i.e. from the proto-experiment, has its origin in the philosophical circle in Miletus in Asia Minor from 7th till 5th century B.C. Conceptualization of such a world-view requires delineation of the system based on categories. These categories, which are also the main categories of the modern chemistry, are *principle* (*element*), *structure*, *change* (*dynamics*), and *preparation*, i.e. the intentional transformation of the nature, the act of creativity in the shaping of new substances. In contrast to the first three categories, the last one declines from science, and it rather belong to the art, the sort of architecture of microcosm, the art of designing new forms of matter. The role of preparation in chemistry is twofold: it is practical for the supply of new materials for everyday use, and, from the other hand, the mindful chemical synthesis expands the general view on the nature. The specificity of chemistry is, in contrast to other scientific disciplines, that it designs its proper objects of investigation!

Principle or element is the category with two faces. It is the matter which we recognize immediately by sensation, the substance in the vessel, visible and tangible, but it is also the abstraction on the philosophical and scientific level. A long evolution of the cognition of the Universe, from sensation to the abstraction, led finally to science and art. It is important to point out here, that even the modern definition of the chemical element as a *mathematical set of nuclei with the same positive charge* is abstract! It is important to recognize that for instance iron, which could be weld or held in hand is the *elementary stuff*, not an element. If the element already in antiquity appeared as an abstraction, the question that appear is the abstraction of what? This is the question that lead to the duality of the notion of element as an abstraction of the substance, and as an abstraction of the principle. If we contemplate about the world as a collection of principles, the image of the substance is vanishing, and, oppositely, if we are focused on the sensible matter, the idea about principles becomes vague. Such kinds of ambiguities, which have later been expressed in the uncertainty principle, will be actualized through the development of philosophy of science, and also through the development of philosophy of chemistry.

Behind the ancient idea about the elements/principles, there is much wider concept about the division of the Universum in two categories, the immobile

Universum and the mutable Universum, i.e. constant and variable. In metaphysics sa well as in proto-astronomy, such division is manifested as immovable stars and movable spheres of the planets. Since the nature of substance has been viewed as a mirror picture of the sky, than the matter should also have its constant and variable component: constant simple elements/principles *versus* variable and transformable complex substances.

But, there has appeared the question about the number of elements. Is there only one element or principle by which all the nature could be represented, or the number of elements is larger? Interestingly, the number of elements has expanded through the history. **Thales of Miletus** (cca. 624–547 B.C.) and **Anaximander of Miletus** (610–546 B.C.) have considered water as a sole element. While for **Anaximenes of Miletus** (cca. 528 B.C.) the sole principle is air, for **Heraclitus of Efez** (535–475 B.C.) it is fire. Reducibility of all the things to the One, proto-mater or proto-origin of the everything is in principle the idea about the unity of Universum, the concept that has been developed in the philosophical system called *monism*. Accordingly, Thales derives all the substance and all the principles from the *prime mater*, απειρον, that in the later history of chemical philosophy appears as PRIMA MATERIA. Such consideration of the reducibility on the One was especially fruitful in the development of chemistry as an experimental science. Is it possible to prepare a sort of prime mater, or at least to approach to it by practical activity, by experiments? This challenge is a Holy Graal of chemistry that has triggered the work of all the future scientists.

Replacement of Thales' water for the Anaximander's humidity, or for the Anaximen's air, is nothing more but the question of representation in the form of different metaphors. However, fire, the element that Heraclitus is setting in the origin of all the things is of different nature. Fire is not only a principle, it is a representation of *change*, *transformation*, *dynamics* and the *irreversibility*. By using fire as a principle Heraclitus departs from the concept of the substance as a basis of element: the world consists not from objects but rather from changes in the vortex of events in the time. Similar philosophical system appeared in twentieth century by Alfred North Whitehead who has proposed that the *elementary particle* should be replaced by the *elementary event*. The nature of fire is not only temporal rather than material, it is also the basic force that can provoke all the other observable changes. In Heraclitus thoughts, element and dynamics are intertwined on a specific way. The intention of the philosophy to find the way, and to approach the elusive point in which the PRIMA MATERIA is concentrated, has provoked the formation of the more and more complex ideas about the elements. The substance with its highest level of complexity is one pole of of the Universum, which represent a contrast to another pole of the distant geometrical chilling of the space. The World seems as a sort of the universal organism, which **Empedocles of Akragas (Agrigento)** (495–435 B.C.) has imagined to be composed from four elements-principles. To water of Thales, to air of Anaximenes and to the fire of Heraclitus Empedocles has attached something stable and persistent—the earth. These four Empedocles' elements, *earth*, *water*, *air*, and *fire* have shaped the schematic concept of the structure of matter, which was actual in the philosophy and science for the following two thousand years. Al of those will be discussed in the upcoming lectures. It is almost impossible to find the reasons that have stimulated Empedocles to propose such a concept of the four elements, because the process of the scientific and artistic invention are elusive and irrational, often they are the consequences of the unaspectable circumstances. Though, in the Empedoclean scheme one could recognize the geocentristic representation of the Universum, with the Earth whose mass is concentrated in its centre, and water and air which form the outer spheres. Already in this time, fire has been traditionally positioned either in the most distant spheres, or in the Sun that was regarded as one of its sources. Long-standing stability of the theory about the four elements could be explained by the fact that its main promotor was **Aristotle of Stagira** (384–322 B.C.) who has constructed almost complete natural philosophy on this Empedoclean concept, and whose authority was unquestionable in the scholastic period of the history of science.

Utterly abstract representation of four elements in the form of perfect geometric solids has been proposed by **Plato of Athens** (427–347 B.C.), who was the long-time teacher of Aristotle, but also his opponent in natural-philosophical questions. Fire was represented by tetrahedron, earth by cube, air by octahedron, and water by icosahedron. By Plato, the ideas expressed as geometrical solids precede the substance. Consequence of this concept was that the geometrical properties of these representations could be applied in explanations of the nature of elements. For instance, the planes of all the platonic solids can be simply divided in particular number of smaller triangles:

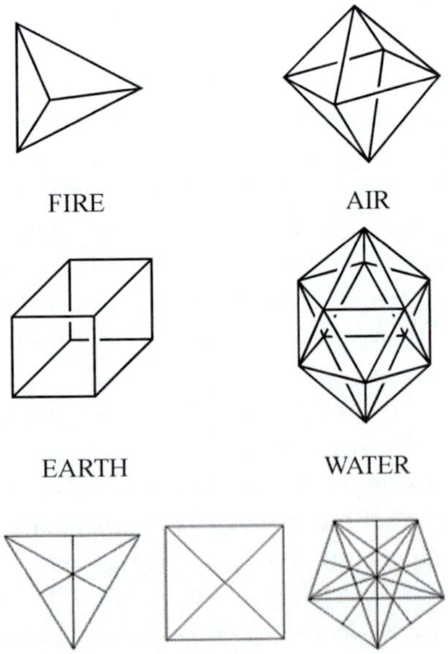


**Plato's representation of four elements as four geometrical solids. The planes of all these solids can be divided in triangles**

Such a way, the triangle appears as an ultimate elemental unit, a sort of *proto-element*, or *proto-atom*. If any of the elements, i.e. its geometrical representation, can be decomposed in triangles, then from these basic units it is possible to reconstruct any other element. Consequently, the elements can transmutate to each other! By this mental experiment Plato intend to demonstrate that the *transmutation* of elements appears first on the abstract level in the world of ideas, and the transmutation of real substances appears only as an emanation of these ideas.

Elements have their own properties, those that we recognize by our senses such as colour, moisture, taste, heat etc. By Aristotle, there are four basic properties of elements: dry, wet, warm, and cold. Every element is characterized by the combination of two properties, which are bound in pairs by opposites: the pair of moisture, dry-wet, and the pair of warmness, warm-cold. Such a way, fire takes *dry* from the first pair, and *warm* from the second, and air takes warm from the first pair and wet from the second.

These pairs of properties, which by their combination form particular elements, correspond to the different and independent types of senses: the sense of heat, and the sense of moisture. As it was already mentioned, none of these two types can be quantitatively analysed and interpreted by mathematics, or at least this was not able in the early history of science. In contrast, other two senses, senses of space and time, can be easily represented by mathematical methods. Movements of the heavenly bodies, for instance, could be calculated and represented as a combination of space and time coordinates.

Heat and moisture can be represented by using a sort of "qualitative coordinates". Every of these coordinates has only two states, and their combination define four elements:

| | "Coordinate of moisture" | | |
|---|---|---|---|
| "Coordinate of Heat" | | Dry | Wet |
| | Hot | **Fire** | **Air** |
| | Cold | **Earth** | **Water** |

Consequently, since the combination of "coordinates" is arbitrary, it is possible not only to generate an element, but also to transmute one element to another. This philosophical concept is the beginning of the idea about the transmutation of elements, the idea that has been for a long time of scientific history in focus of chemistry and alchemy, till the seventeenth century. The idea of transmutation has been extended during the history to one of the basic categories in chemistry. In principle, all substances could be transformed in another stuff especially if this transformation occurs not simply, but through more steps of which the first step must be the preparation of the simplest substance, the *Materia prima*. Since, by Aristotle, real substances are various combinations of the four elements in differing ratios, the principle of transmutation/transformation is valid also for complex materials. Accordingly, one substance is transformed in another merely by changing the ratio of elements. Both, Aristotle as well as Plato afford that the four elements are immerged into the fifth principle, the *Quintessence* or *Aether*, which perennially rests above the terrestrial sublunary world.

As always in the history of science and philosophy, every idea triggers the appearance of its contrary. Thus, the concept of element had a counter idea: maybe there are no elements, and no transformations at all. For **Anaxagoras of Clazomenae** (510–428 B.C.) elements are unnecessary categories, there is only an initial sense of the world, the *nous* (νους). This concept is similar to the *logos* (λογος), the primordial word, the initial structure, and the first step in the creation of the Universe. By Anaxagoras, the Universum emerges from the eternal quietude by an *action of the disturbation*, the *spermata*. Spontaneous fluctuations are indeed among the basic a priori heuristic ideas also in modern science. For **Parmenides of Elea** (nearly 515–450 B.C.) there are no changes, no transformations. There is only a deep essence of all things, their Being, the *onthos*, οντος. As we know, the investigation of the essence of the substance, the *ontology*, is in the core of science.

On the other hand, the *structure* as a pattern by which the world could be recognized and analysed in its various appearances, belongs to higher levels of complexity. In principle, there are two viewpoints on the structure of matter: the substance is continuous, and the substance is discrete, i.e. discontinuous, composed from the arbitrary number of some basic units.

The problem of discrete/continuous structure of matter is one of the most vigorously discussed concept in science and philosophy that lasts untill the modern times. By Anaxagoras, the substance is continuous and infinitely divisible. The same idea in its radical form has advocated Aristotle in his lectures in Athens' **Lykeion** (Λυκειον, established 335 B.C.). Aristotle affords:

> Things are continuous when any of two neighbours on their boundary become one body, and, as follows, they are in conjunction.

Thus, the substance is fulfilled and indivisible, there is no empty space and no separated entities. All is melted in each other. Aristotelean idea about the structure remains unclear; it seems that the substance consists from two components, **stuff**, hyle (υλη), and **properties**, which the substance can obtain only if a being, ουσια, is added to it. *Such a concept in which the properties are considered as the categories separated from the substance has for a long time been in focus of chemical philosophy.*

The idea about the *discrete*, discontinuous structure of the world is as old as the idea about the continuous structure. Natural and rational numbers as the abstract building blocks of the world were in the basis of the doctrine of **Pythagoras** (582–496 B.C.). Though the discrete structure of the matter has explicitly proposed by **Democritus of Abdera** (460–370 B.C.), and **Leucippus of Miletus** (first half of 5 Century B.C.). By their opinion, the substance cannot be divided *ad infinitum*, but only till the smallest basic particles, the atoms (ατομος), which are characterized by some elemental attributes related to the macroscopic sensuous properties of the substances. Atoms have their shape, position and arrangement, and these properties are mirrored in the properties of visible substances. Since the arrangement is one of most important characteristics, the observable substances are reflections of aggregates of atoms. From the modern point of view, Democritian atoms are similar to molecules, rather than to atoms. It is clear that the idea about atoms includes also the existence of the empty space, in contrast to the concept of the fulfilled space, which arises from the presumption of the continuity. Polarization on these two opposed comprehensions of the space is still an actual problem in science and metaphysics, and it appears as one of the basic questions in cosmology.

From the metaphysical point of view the problem of matter, the substance, is in close relation with the concept of *identity*. The metaphysics of identity Aristotle is examining by unification of two notions, *subject*, *hypoikeimenon* (υποκειμενον), and *being* (ουσια).

Later in this book I will discuss two levels of identity: the *primary identity* that represents the identity of elemental microparticle, and the *secondary identity*, which is the identity of the particular unit or individual (particular macromolecular structure, or living cell).

Since Aristotle in his physics does not accept empty space, he did not consider the existence of space as a category. Instead the space, Aristotle is using the notion of *position* (τοποσ). Fundamental category in his physics is the *time* in the sense of *change*. Focusing on υποκειμενον, μορφη, and στερεσις (*subject, form, and deprivation*), i.e. on *substance*, *properties*, and the *possibility of change*, has been the ideal starting point for the development of chemistry—the practice of mutual transformations of substances! Such Aristotelian view has probably been the reason that the space as category did not appear in the traditional chemical science until the second half of 19th century, when **van't Hoff** and **LeBel** have established the basic concepts of spatial configurations of molecules.

Aristotelian words without the space on the one side, and the atomistic Platonian world based clearly on the analysis of a perfect aprioristic space on the other side, both were reflected through the long history, not only in science, but also in other fields such as fine art. Mediaeval sacral painting, especially after the philosophy of **St. Thomas Aquinas** (1225–1274), and the establishment of the authoritarian aristotelism is not focused on the representation of space but rather on the representation of the hierarchical structure of the world. Human figures are planary positioned on the (frequently gold) background. In contrary, Renaissance that emerged from the resurrected Neoplatonism has introduced another comprehension and representation of the space by using perspective. The figure by mediaeval gothic painter **Cimabue** (nearly 1240–1302), and the renaissance figure by **Fra Angelico** (nearly 1395–1455) that is shown below it clearly demonstrate this conceptual change.

**Cimabue:** ***Maestà di Santa Trinità,*** **1280–1285, Galleria Uffizi, Firenca**

**Beato Angelico: L'*Annunciazione del corridoio Nord,* 1440–1450, Museo nazionale di San Marco, Firenze**

Contrary to the Aristotelian physics and metaphysics, Democritean and Leukipian viewpoints about atoms comprise also the first ideas about the structure, configuration, and proto-shapes, the concepts which have reached its culmination in the formation of the structural theory in the nineteenth century. However, we have to keep in mind that both notions, element, and atom are in principle abstract. Reality and materiality of atom was not completely accepted until the beginning of the twentieth century.

**Epicurus of Samos** (341–270 B.C.) has extended the concept about the discrete, discontinue nature of the matter by adding two ideas for which we could tell that they belong to the core of modern science. Besides the material, Epicurus introduce also temporal atoms, say, time is also discontinuous! Interestingly, the definition of atoms as the smallest units of time appeared in scientific and philosophical circles until seventeen century. In his Dictionary of Croatian Language, ***Gazophylacium***, written in seventeen Century, **Ivan Belostenec** (Bilostinec), (Varaždin, 1593–Lepoglava 1675, Croatia) describes atom as the smallest particle of time:

> ATOMUS, 1. *Prah, y iszkricze one drobne, koje po trakeh szuncheneh lechu, gda szuncze prek obloka szveti* 2. *Naymenyssi* ***hip vremena, kisze namenye razdeliti nemre***
>
> (1. *pulwer, and small sparkles which are visible on the Sun rays shining through the window;* 2. ***the smallest moment of time that cannot be divided***;)

Another Epicurus' idea was also of a far-reaching importance: the movement of atoms is always from up to down, and they accidentally can decline from its trajectory. Such a way, Epicurus has introduced *contingency without the cause as an elemental event.* Statistical nature of the Universe, which is one of the basic concepts of modern science, is perhaps the most revolutionary Epicurean invention. Of course, this construct appeared too early in the history, since the Aristotelian *causal principle* has predominated in the natural science and philosophy until the last quarter of nineteenth Century. Florentine Neoplatonistic philosopher **Marsilio Ficino** (1433–1499), in spite to the notion that he was an establisher of Renaissance, has in his works about astrology strongly opposed to the epicurean ideas on contingency in the nature. Epicurus' successor in the philosophy about atoms was roman naturalist **Titus Lucretius Carus** (first Century B.C.), who, in his work *De rerum natura*, has continued in the explanation of the structure of matter as discrete. Most of samples of his work were destroyed in the time of early middle ages, undoubtedly because of its inopportune aspects. Uncertainty about the *causality* versus *contingency* was in natural philosophy transformed in notions that we know as *determinism* and *indeterminism.* In works of Marsilio Ficino it can clearly be recognized his attempt to reconcile determinism based on the law of causality, and indeterminism where the interpretation of the world is statistical. By Ficino, an absolute determinism and immobility is characteristics of the Universum in its highest spheres. In lower levels of souls some movements are possible. By approaching from Heaven to Earth and to the sublunary world, the movements are more expressed, and, finally, on the level of living beings, which are most distant from the divine primordial cause, the influence of the immediate environment overcomes the impact of God, and, consequently, contingency is predominating [1].

Within the framework of our conjectures about the proto-origins of physic and chemistry, based on different categories of senses, the proto-physics, which has been emerged from astronomy and metrology of land could only be deterministic. On the other hand, ancient chemistry, with its uncertain experiments and observations, was indeterministic: every trial was unpredictable. Obviously, chemistry and physics, both were developed as two complemental concepts also in this sense.

Chemical indeterminism could be recognized also through typical chemical category—the change as a fundamental nature of substance. This category of transformation contributes to all other chemical categories. As we have already mentioned, transformation is directly related to considerations about elements, especially in the philosophy of Heraclitus where the fire is a basic element. In contrast to water, air, and earth, fire is an elusive principle of change. The event of fire is *irreversible*, in the same way as it is every living system far from any equilibrium. The change has its beginning and its end, and a world is nothing more than juxtaposition of changes, rather than composition of objects. This philosophy and metaphysics has been resurrected in the first half of last century by **Alfred North Whitehead** [2]. Aristotelian transmutations of elements include the change, κινησις (kinesis), and movement αλλοιωσις (aloisis, aloisia) in the focus of speculations about substance. Ancient chemical ideas have their origin in the Aristotelian discussion about *creation and disappearance*. New created substance could be either composition συνθεσις (synthesis) the mixture, or μιξις (mixis), the substance which we know as a chemical compound. The statement that the interaction between the substances is possible only if one of them is active, and the other is passive, is the first antecedent of the chemical dualism developed in nineteenth century by **Jöns Jacob Berzelius** (1779–1848) who afforded that only different atoms can be bound to each other. In the Platonic philosophy the origins of changes are discussed more generally, as consequences of inhomogeneity.

*Experiment* as the exclusive approach to explanation of elements, structure, and dynamics of matter appears in the antiquity only in its undefined forms. This method did not arise from philosophy, but from everyday practice, from handicrafts as well as from the mystique of rituals. Smiths and physicians were among the closest to experiment. Preparation of metals in furnaces of smitheries and foundries has been considered as a mimic of occurrences that appear in the Earth's interior, rather than in heavens. Furnace or vessel, as Earth in small, were in the focus of interest. Everything is occurring in the centre of some instrument, *in medio centri*, but the results of most of the processes were uncertain. To obtain the expected substance, metal, or medicine, the trial, the mastery, have to be repeated many times. Observations were focused on colour, heat, volatility, solubility, softness or taste. Evidently, here is no place for determinism, arithmetic, or geometry.

## 2.1 Conclusions

The appearance of particular sciences, such as physics and chemistry, has its anthropological origin in the nature of the sense data. Physics has emerged from the senses of time and space, the categories that can be measured and analyzed mathematically. Consequently, this science has developed the general theoretical representation of the Universe. In contrast, chemistry has appeared from the sense data of color, smell or heat, the data which were at the time not measurable. Rather, these sense data are related to the real practical manipulation with the substance. Chemistry has appeared as an experimental practice which can immediately transform the matter. Concepts about the universe, which emerged from such manipulation with substance are of general and philosophical importance: element, principle, and change.

## References

1. Ficino M (1999) Scritti sull'astrologia. RCS Libri, S.p.A, Milano
2. Whitehead AN (1978) Process and reality. The Free Press, New York

# 3 Synthesis of Philosophy and Technology

*All is interrelated. Heaven and earth, air and water. All are but one thing not four, not two and not three, but one. Where they are not together, there is only an incomplete piece*

Paracelsus

**Abstract**

This chapter is dedicated to the first appearance of chemistry, the Alexandrian proto-chemistry. The practice of experimentation with its scope to materialize the Aristotelian, and Platonian philosophies, which has appeared during the period of Hellenism, is considered as an origin of chemistry, and as the predecessor of Arabian and European alchemy.

**Keywords**

Alexandria · Hellenism · Proto-chemistry · Experiment · Philosophy of change · Philosophical gold · Great Work · Opus magnum · Aurifictors · Aurifactors

Ladies and gentlemen,

Two origins from which chemistry has developed, philosophy with its basic categories (element or principle, continuity, discontinuity, transformation, structure), as well as craftsmen practice of smiths and metallurgists, got together in Alexandria, in the new city that has been constructed in 331 year B.C. New science, the new paradigm, emerges in the history almost always as a result of the synthesis of often quite different practices and ways of thinking. From the analysis of the historical conditions in which the most important theories were invented, we could draw the rule by which a scientific revolution, the change of paradigm, results from amalgamation of concepts from different fields. Thus, the scientist has greater chance for


H. Vančik, *Philosophy of Chemistry*, Integrated Science 2,
https://doi.org/10.1007/978-3-030-69224-7_3

important discovery, if he cognizes and examine ideas and concepts of other scientists from as wide as possible spectrum of research. Solutions of contradictions that could appear in a system are not possible within this system, but rather by a courageous step out the system, the formation of the view from outside and its possible extension. A scientific lifetime exercise focused to one stuff, and studying only the papers about this stuff, does not lead to new conceptual and intellectual adventures.

The project of **Alexander the Great** (356–323 B.C.), who has been well educated under Aristotle's supervision, to establish a city and civilization on principles of a synthesis of various views and thinking was a longstanding idea. Museum (Μουσειον—temple of Muses), the central Alexandrian educational institution, was the melting pot where thinkers from all world of that time, Jewish, Greeks, Egyptians, Babylonians, as well as thinkers from the Far East met together. Greek philosophy of Plato and Aristotle, and the advanced Egyptian metallurgy and the craft of preparation of colours and jewels were somehow mixed with the Persian mysticism. ***Protochemistry***, which in its base had an experiment as a new philosophical method, has appeared as a new discipline [1, 2]. This new approach to the knowledge was epistemologically deductive: the theory of current philosophical systems, which originate from thinking rather than from hard experience should be submitted to test. Experiment is expected to be the method to demonstrate the extent of Aristotelian and Platonian constructions, and to support them or to reject them. Aristotle's statement about the intention of all things to occupy their natural position has been generalized in way that all substances and objects tend to obtain their perfect form. Perfect substance, brilliant and resistant—the philosophical gold that is "golder" than the ordinary gold, and which is useful for the perfection of all other substances, could probably be prepared by a patient repetition of experiments. Manipulation with materials, which led to the development of the future laboratory equipment, was at that time imagined to have some connection with the cosmic principles. Platonian and neoplatonian construction about spirits captured in the body, which should be liberated, was a philosophical base of the establishment and deliberation of two categories: the *shapeless matter*, and the corresponding *feature of matter*. Substances appear as noticeable only when they are unified with the features! Accordingly, the properties became the category independent on the substance. The role of experiment is to move properties from one object to another. In laboratory, the Master manipulates only with features: metals or alloys that possesses an extent of some feature could this extent transmit to other substances for their perfection. The *philosophical gold* is in principle a sort of "supragold", and some fraction of its pool of goldness could be transferred to the shapeless matter to make it excellent. In addition, the philosophical gold is timeless, and this symbolism we can recognize as a gold background in icons of the Eastern Christianity: Christ and Madonna are immersed in gold!

Features are spiritual, the *pneumas*, and they could be released from the substance by sublimation, and added to another substance. All the process is observable: for instance, if the feature produced by evaporation of mercury is transferred to copper, we can observe change of its colour. Thus, the effect is recognizable by the perception of the change of colour—the sense datum that is not possible to analyse mathematically. Indeed, colours as basic manifestations of features became in the focus of interest in protochemistry. The transfer of colour from one shapeless matter to another can not be performed arbitrary, but rather by respecting the strong order and procedure. In antiquity, the colours were ordered hierarchically: from black, to white, to yellow, and to the perfect purple. **Comarius** (1st century), one of the earliest protochemists, describe blacking, whitening, yellowing, and reading as a basic steps in the performance of the **Great Work (OPUS MAGNUM)**. This procedure has appeared as the central method of ancient chemists. None of these steps cannot be skipped! Since the black colour is a representation of the primary matter, the procedure should always begin with blacking.

The initial version of the idea about levels of complexity, which was proposed in first lecture in this book, has appeared as the four levels in the practical realization of the procedure of OPUS MAGNUM. The occurrences on the higher level are not possible and not completely determined exclusively on the basis of the structure of the lower level. (as, for the instance the biological event cannot be derived exclusively from chemical phenomena and laws). As we have seen, Marsilio Ficino made the further step in the development of the mataphysics of levels of complexity in the time of Renaissance. Let us live this discussion to the lectures later in this book. At the moment, it is enough to show that the systematization of features has anticipated the systematization and classification of substances. Protochemical experiment is an invention, the design of the method by which it would be possible to manipulate with the elusive senses of colour, heat, humidity, solubility, or taste, and which had a hidden goal to test the Platonian and Aristotelian metaphysics by deductive experimental methods. If Karl Popper would use chemistry as origin for his philosophy, he would find excellent examples for the concept of the testability as a main characteristic of a serious (proto)science.

For protochemists it was necessary to find the relationship of abstract ideas such as four elements, logos, or primary matter with the real substances, metals, minerals, liquids, the substances that origin from living beings. Mental constructions ought to be closely related with practical manipulation with substances. Devices invented for such a practice were a sort of "mutated" craftsmen tools, and the processes that have been performed on the controlled way were from their practical goals transformed into the methods of representations of the abstract metaphysical concepts. Such a way, gold has changed its function from decorative to philosophical, and it became the substance that is, because of its perfection, the closest to the *perfect materia* i.e. to the origin of everything visible and perceptible. Producers of philosophical gold, the *aurifactors*, became therefore the first chemists with whom the long way to the science called chemistry had begun. The etymology of the word *chemistry* is still under discussion. Some origins of this term are known: the word *kheme*, *khem* or *khmi*—that means *black earth* or *Egypt* in the old Egyptian language, or *chyma*, as old Greeks call the *melt*, or *kiem* in Chinese in meaning *gold*. The most convincing is the word χημεια (*chemeia*), which had been found in the works of **Plutarch** (46–120), what is the name for the ancient skill of the preparation of precious metals and manipulation with substances [1].

Since the uncertainty of the results of experiments in protochemistry was quite contrary to the certainty of the predictions of events on the sky in protophysics, the language for the description and explanation of experiments had to be accommodated to this new "ambivalent" discipline. While the language, say scientific language, of astronomy and geometry was already almost defined and accepted because it was based on the mathematized approach to the space and time, the language of protochemistry was, in contrary, quite vague, ambiguous, and individual. In contrast to the ancient astronomers, say, protophysicists, which had a sort of scientific language understandable for all, the protochemists were individuals in that. Their experiments were indeterministic, and the regularities of particular experimental observations were in principle statistical, they appeared only after a series of repetitions. Scientific language for the representation of the indeterministic experiments should have the same nature—it must also be indeterministic. It must be a secret, an Arcanic language, which was often understandable only to the author of the text. Moreover, protochemists even intended to make their language as vague as possible, because the idea of the public rapport about their experiments and reflections was unacceptable for them. Protochemists not only disregarded their successors, but they considered them as undesirable. In the chapters that follow I will show how the protochemical language in its arcanic structure has reached a peak in the later European alchemy. A narrow connection of the language and practice, in the early time of the secret experiments and secret language, played a key role in the formation of almost all chemical theories. **Interweave of the syntax of the chemical language, as well as the quantifications of observations of the substance transformations were in the core from which the structural theory in 19th century has been constructed**.

The sublunary world as a reflection of the events in heaven was present in the symbolism that protochemists have used for the representation of metals [3]. The symbols for metals and the symbols for heavenly bodies were the same. That means that both the symbols are nothing else but the reflection of some higher abstract spiritual structure. Seven heavenly bodies, five planets, Sun, and Moon were in analogy with the seven known metals. In a remarkable literary text by **Bela Hanvas** (1897–1968) it can be find that **Adamantius Origen** (184/5–253/4), the professor at the Alexandrian Museum, has presented his lectures simultaneously in seven languages because he intended to demonstrate the deeper meaning of abstract concepts. Historically, it is known from the documents, that Origen had probably the highest influence on the emergence of the six lingual Bible (‘Εξαπλα, Hexapla, six-fould) in which the texts on different languages were represented in six parallel columns. The **method of analogy** that is apparent in this symbolism appeared as more and more powerful contrast to **the method of logics**. While physicists in their analysis of the structure of the World have a support in logics, chemists are contemplating by using analogy. As we will see later, analogy is a sort of the qualitative category. This epistemological polarization in logics and analogy is one of the arguments that support our thesis about the different origins of chemistry and physics.

Experiment, philosophy, and mysticism were in different ratios entangled in the works of early protochemists [2]. Two of them have used the name of Democritus.

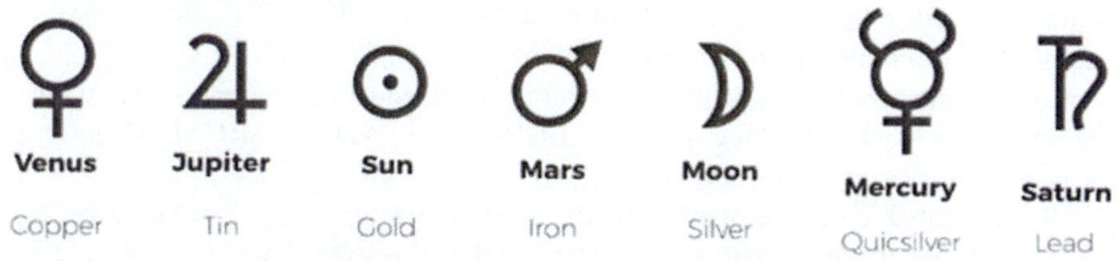


**Protochemical symbols of metals and their planetary analogues**

**Bolus of Mendes** (3rd century B.C.), who is in literature known as **Pseudo-Democritus**, was especially important for protochemistry. First writings about substances, medicine practice and magic origin from him. More details are known about **Maria** the **Jewess** and **Cleopatra** [3, 4] (1st century). Mary Jewish has invented a series of laboratory procedures and instruments, *inter alia* the *kerotakis* (κηροτακις), the device that has been a combination of sublimer and extractor. The solid substances were heated and vaporized from the bottom of the vessel, and corresponding vapours have passed through the grating and finally condensed, solidified. Evidently, vapour, or *pneuma* is imagined as a sort of *feature* that evolved from the substance on the bottom, and which is then transferred by condensation to another substance, the metal on the upper grid. The observed colour changes were used as an argument that supports the concept about transfer of features between different substances. Similar procedure of evaporization of the pneuma-vapour from liquid and its condensation in another vessel was distillation. This is one of the oldest laboratory procedure, which is with minor variations also used today. The apparatus consists from the same main components: vapour condenser (*ambix*), and fraction collector (*bikos*).

Thus, the theoretical background of the first chemistry became complete: the "structure of matter" consists from two parts, the unshaped substance and the feature. By evaporation and re-condensation, the features could optionally be transferred between the black basic substances, and all this procedure is controlled by observation of the colour change.

Continuous repetition of procedures, and focusing on these cycles of experiments together with the intellectual immersion in this metaphysical world did not live chemist to be personally indifferent. **Zosimos of Papanopolis** (350–420) was one of the first protochemists who has attached the significance of the experimental performance to the personal psychological and mystical experience. His experience of changes of substances, Zosimos has associated with the personal adventures in the dreams. The protochemical procedures Zosimos has described as parables which reflects spiritual metamorphosis: *nigredo*—the obscuration to reach the primordial state, followed with enlightenment. Transformation, conversion that appeared as a central event in Christian theology, philosophy and ritual has been related to the practical chemical labour: HOC EST CORPUS MEUM—*This is my body* became HOCUS-POCUS on a popular level. Conversion should begin with blackening, i.e. the return to the unshaped proto-substance that has been symbolized either as lead on the Earth, or as Saturn on the heaven. Saturn, the Melancholia, is an origin from which it is possible to re-shape new enlighten substance. Each such act of enlightenment must follow the same order and procedure: to start with the return to void, to the beginning—REGRESSUS AD UTERUM. Crucifixion, decease, and Resurrection!

It seems that protochemistry already in its appearance has developed two tendencies, *exoteric*, *operative*, i.e. experimental-philosophical on the one side, and *esoteric*, *speculative* that was apparent in Zosim's thinking on the other side. This second tendency has almost entirely dominated in the medieval European alchemy.

Not only today, but also during all the history, science and philosophy were endangered from pseudo-science. In such a way, protochemists which found their interest in philosophical gold, and which were known as *aurifactors*, have get their public concurrence in *aurifictors*, the manipulators which have produced a fake gold for a black market. As the nature of contemporary pseudo-science is eclectic, i.e. it combines anything with anything, for instance quantum mechanics with feminism and pets, just to be well sold on the market, so the Alexandrian aurifactors have referred to the philosophy of protochemistry to sell fake gold. These sorts of "gold" were in principle alloys with some minor percentage of gold. We have to know that at the time there were not methods by which it would be possible to recognize unequivocally the real gold from the fake gold. Since ancient Roman administration did not accept the production of gold without their supervision, they have decided pragmatically to prohibit both, aurifictors as well as aurifactors. They did not recognize the difference between them and, consequently, protochemistry became out of the law. However, expulsion of protochemists from the Roman Empire had also other reasons. After the emperor Constantinus, the official religion in Rome became Christianity, which has in this way been transformed from its basic idea of the Profeta-Illuminator to the pragmatic hierarchically organized ideology. In 4th century bishop **Theophilius** finally have burned away the Alexandrian library, and Hellenism became out of the law. In the year 529 the emperor **Iustinianus** has finally abolished Platonian Academy in Athens.

Basic metaphysical system developed within the protochemical circle was logically incoherent: the changes and transformations of substances have been explained as a combination of two kinds of transmutations. Protochemists have accepted the Empedoclean division of the world on four principles-elements, as well as the Aristotelian idea about their transmutation based on the combinations of properties such as wet-dry and warm-cool, while the transmutation of metals they explain as a transfer of free features. However, which was the relationship between features and properties? Vapour, the pneuma, which has been observed experimentally in apparatus for distillation or in kerotakis could tentatively be categorized as feature. But, in accordance with Aristotle, the same vapour is nothing else than a combination of properties dry-could. Does it mean that all the features are dry and could? Does it follow that feature consists from two properties?

From this contradiction it follows that *logics is not decisive* in protochemical theory. Contradictions about which the old chemists were conscious were overcomed by a quite different independent approach—*the use of analogy that can be successfully applied in the definition of* ***quality***. By their «orthogonal» position to the logics, protochemists have established new epistemology, which, through the history, led to the basic knowledge about the microstructure of matter. The reasoning in analogies supposes the use of a referent visible property, again, the quality. For instance, two substances could be considered as analogous if both builds the same shape of crystals. In this case, the referent property is the shape. In investigations of transmutations of metals the referent property was colour. Thus, the transmutations that exchange the same colour are analogous. All "golden" is gold! It is almost unimaginable that the science such as chemistry has developed from logical contradictions and from the provisory such as analogy. Yet, the basic chemical concepts, such as the structural theory and the periodic system of elements, have developed mostly by epistemology of analogies!

## 3.1 Conclusions

Although alchemy is in the most literature origins considered as a predecessor of chemistry, in the history of science the activities that led to chemistry in the western civilization appeared much earlier. In the Hellenistic Alexandria, the philosophies of Plato and Aristotle were combined with the practice of craftsmen and with the mystical spiritual experiences of Egypt and East. Aristotelian concept of transformation of substance has found its analogy in practical performance of transformations in metallurgy and in preparation of remedies. Transformations of the substance in that way changed its role from practical to philosophical.

Philosophical experiment is born and the first activity similar to chemistry has appeared—the Alexandrian proto-chemistry. Proto-chemical experiments have manipulated qualities such as color, or heat or melts, the qualities that are not measurable, in the contrast with the proto-physical theoretical concepts based on the measurable time and space.

## References


1. Grdenić D (2001) Povijest kemije, Liber, Zagreb (History of Chemistry, original in Croatian)
2. van Brakel J (2000) Philosophy of chemistry: between the manifest and the scientific image. Leuven University Press, p 160
3. Greenberg AD (2000) Chemical history tour, picturing chemistry from alchemy to modern molecular science. Wiley-Interscience, New York
4. Brock W (1992) The fontana history of chemistry. Fontana Press, London, 1992, p 16

# Alchemy

# 4

*Whosoever has made me white also makes me red. Both white and red spring from a single root.*

Arnold of Villanova


**Abstract**

Here I argue how alchemy has its origins in proto-chemistry, and which new developments in this period of chemical philosophy gradually led to the development of chemistry as science. Philosophical aspects of Alexandrian proto-chemistry were transformed into more practical aspects of (al)chemistry: their use in medical practice. The concept about four elements taken from the classical philosophy has been extended with the three new principles, and with new abstract concepts that are in the core of modern science. It is explained how alchemy with all its basic principles was transferred to Europe. Problem of authorship of scientific and philosophical works in that time and later in Western culture is also discussed.



**Keywords**

Alchemy · Arabian alchemy · Principles of salt · Mercury and sulfur · Hermes Trismegistos · Esoterism · Exsoterism · Subject-object problem · Iatrochemistry


Taking into account the fact that the collection of sources and works about alchemy is almost immeasurable, I am conscious, ladies and gentlemen, my colleagues, about the risk to which I am exposed by talking about the theme that is imposed as unavoidable. The word *chemistry* or *chemea*, or *chema* receives in Arabian language the grammatical article: thus *al-chemy*. As we can conclude from previous lectures, this word was erroneously used to name the first ancient chemistry, even






H. Vančik, *Philosophy of Chemistry*, Integrated Science 2,
https://doi.org/10.1007/978-3-030-69224-7_4

within the educated circles. Prosecution of protochemists from Roman Empire, about which write historians, has finished in Persia. The elite of Hellenistic culture and science has moved to the old Persian university Gondishapur, which, together with almost the entire territory of Middle East, North Africa and todays Spain, became a part of Arabian region constituency. Arabs were not only tolerant to new science, but they have continued with its further development. Arabs have established universities in Cordoba, Toledo, and Granada that became also the centres of European culture.

Hellenistic and classical scriptures were translated in Arabian, but with some modifications and additions. The dominant doctrine has been Aristotelian metaphysics about the stationary supralunary and the variable sublunary world, which was in harmony with the Islamic idea by which the events on Earth are just a reflection of events in the heaven. Perhaps, this idea has stimulated the vigorous development of astronomy in this period. The studies of occurrences in heaven have to be applied for the explanations of the terrestrial occasions: astronomy because of its application in the astrology! **Abdulah Ibn-Sina** (980–1037, in Europe known as **Avicena**, and **Ibn-Rujd** (1126–1198) with the European name **Averroes** have explained the hierarchical structure of complexity of the Universum on the new way. Alchemy was readily accepted as one of the independent sciences, as it follows from the work *Turba Philosophorum* by **al-Ihmimi**, the author from 11th Century. The term *kimia* as the skill and science about transformation of metals can be find in the book from 9th century associated to **Abu Jusuf Jakub ibn Ishak ibn al-Sabah** (801–873), who is in European scripts known as **Alkindus**. His chemical work covers preparation of perfumes and the skill of distillation but not the science about metals [1, 2]. Perhaps, Alkindus was one of rare chemists who did not accept the concept about the transmutation of metals.

One of practical problems in the history of science, and in history generally is the question about authorship. In the early history, the philosophers and writers did not subscribe their works with their own names. Rather, they have used the names of famous authors. Let us remember Bolos of Mendes, who has subscribed his works with Democritus. Explanation by which the name of famous author has guaranteed the quality of work is only partially true. I am inclined to favour the opinion by which the name of author has not been important in antiquity because everything important on the Earth and on the heaven was known from ever. In her script, *Gold Making of Cleopatra* its author wrote: *One is All and through it is All, and by it is All, and if you have not All, All is Nothing.*

This is a wisdom that is from ever to ever. What about authorship in this?

This *everything and from ever* has in the philosophy transformed in the personalized form of proto-spirit of all the sages–**Hermes Trismegistos**. This legendary father of all chemists, whose trace has been vanished in the historical fog, have appeared as an author of the learnings by which *everything is known from ever*. His *Tabula Smaragdina* (Smaragdine tablet), of which the oldest available version is the 12th century Arabian translation, became a cannon of all alchemists, the summary of all the knowledge. Let us quote one of its translations by Isaac Newton:

*Tis true without lying, certain and most true.*

*That which is below is like that which is above*
*and that which is above is like that which is below*
*to do the miracles of one only thing*

*And as all things have been and arose from one by the mediation of one:*
*so all things have their birth from this one thing by adaptation.*

*The Sun is its father,*
*the moon its mother,*
*the wind hath carried it in its belly,*
*the earth is its nurse.*

*The father of all perfection in the whole world is here.*
*Its force or power is entire if it be converted into earth.*

*Separate thou the earth from the fire,*
*the subtle from the gross*
*sweetly with great industry.*

*It ascends from the earth to the heaven*
*and again it descends to the earth*
*and receives the force of things superior and inferior.*

*By this means you shall have the glory of the whole world*
*and thereby all obscurity shall fly from you.*
*Its force is above all force,*
*for it vanquishes every subtle thing and penetrates every solid thing.*
*So was the world created.*

*From this are and do come admirable adaptations*
*where of the means is here in this.*

*Hence I am called Hermes Trismegist [sic],*
*having the three parts of the philosophy of the whole world*
*That which I have said of the operation of the Sun is accomplished and ended.*

*Hermeticism*, which is relying on Hermes Trismegistos, has appeared as a spiritual and intellectual movement probably in the 2nd century within the Hellenistic circle, and it has very important influence on the occult component of protochemistry and alchemy. This spiritual movement took various ideas from almost all philosophical schools of that time. However, by statements of some historians of science, hermeticism has not been a coherent philosophical system. Its relationship with protochemistry and alchemy is based on the common practice in the use of language for the explanation of the world. **Umberto Eco** (1932–2016) has suggested that Hermeticians have considered that every text could have indefinite number of interpretations [3]. Perhaps, this was a reason that Origen of Alexandria has lectured in Museum simultaneously "in seven languages". In this frame, *Tabula* could have also indefinite number of interpretations. From the point of view of modern chemist, it is also possible to discover in Tabula a real chemical experimental procedure.

The irrelevance of authorship can clearly be seen from the history of the name of **Jabir, (Abu Musa Jabir ibn Hayyan**, 725–815) to whom the alchemical work

*Summa perfectionis magisterii* has been attributed. Jabir, who was in Europe renamed in **Geber**, has not been the author of this work. Behind the name Geber was the Franciscan **Paul of Taranto** from 13th century (i.e., **Pseudo-Geber**). The *Monistic idea* about the general unity, and that all is known from ever is present in *Tabula Smaragdina*, but its origin remains elusive. Continuous linkage with the common divine wisdom, which successively appeared in Orphism, Hermeticism, Pitagorean philosophy, Zoroastrism, and Christianity has been synthetized in the proto-theology, which the renaissance thinkers **Georgio Gemisto Pletone** (1355–1452), and **Marsilio Ficino** (1433–1499) have named *prisca theologia* [4]. (Ficino, 1999). Within this learning, since everything is known from ever, an authorship is quite unimportant. This is probably the reason that the real authors and declared authors were different. Declared authors of these works, which were in principle the names of famous thinkers, have warranted their importance. Probably this has motivated **Paul of Taranto** to publish his books under the name Geber.

Although Jabir did not played particular role in chemistry, his scripts about the theory of science were important contribution. Jabir has classified sciences in two categories: higher sciences and lower sciences. In the higher "rational" sciences, which Jabir calls also the sciences for "hereafter", belong Aristotelian physics, astronomy, arithmetic, geometry and metaphysic. However, alchemy and medicine were classified as lower, earthly sciences. Already in this classification have appeared the contours of the future structure of the philosophy of science developed in the Vienne Circle in which chemistry was not classified as the proper science. Immanuel Kant, who analogously categorized chemistry in skill rather than in science could find Jabir as his predecessor!

Alexandrian protochemistry and alchemy differ in two aspects. While the scope of protochemistry was philosophy in the first place, alchemy was more practical, its scope has been medication: if it is possible to use practical procedures for the perfection of metals, than the similar procedures could be applied for the perfection of the life itself to attain absolute health, or, probably immortality. The concept of the protochemical philosophical gold has been transformed in the alchemical "stone of wisdom" (LAPIS PHILOSOPHORUM) and *elixir* (al-ixir), the esoteric substance that can be used not only from obtaining health, but also for transmutation of metals to gold.

Second difference between protochemistry and alchemy was conceptual. Behind the four Empedocle's elements, alchemists have added three additional principles: *salt, sulphur, and mercury*. Salt is the principle of persistence, sulphur is the principle of flammability, and mercury of fluidity. This alchemical axiomatics that is composed from four elements and three principles has arisen as a long-standing base of European alchemy.

Although the method of Arabian alchemy was in principle esoteric, in works of some alchemists we can find the roots of the future scientific chemistry. In this line, the especially important personality was **Abu Bakr Muhammad ibn Zakariyya al-Razi** (nearly 865–925), the alchemist who has replaced esoterism with the experimental science focused on the investigations of the nature. In his works, *The Secret* (Al-Asrar), and *Secret over Secrets* (Sirr al-Asrar) al-Razi has developed the method which could already be called scientific. In his books Al-Razi classifies the substances (in herbal, animal, mineral, metals, and non-metals), the laboratory equipment and experimental procedures. He was probably one of the first chemists who has used the quantitative approach by weighting of the material before the experiment, and after the experiment. Al-Razi describes laboratory experiments of the reactions of metal oxides with coal that yielded metal. The term *alkali* probably origin from his works. The name of potassium, *Kalium*, which is present in alkalic solutions has the same etymology. Combustion of substances of herbal origin yields ash, *al-kali* in Arabian.

As the most of the ancient classical knowledge, alchemy was transferred in Europe where it survived in different circumstances and forms until the time of the first scientific revolutions. All the famous thinkers, from the early Middle Ages till the Renaissance, were more or less alchemists. Without alchemy, they did not succeed to harmonize their theological, and philosophical learnings. **St. Thomas Aquinas** (1225–1274), **St. Albertus Magnus** (1200–1280) and **Roger Bacon** (1214–1294), the theologians and philosophers, which have accommodated aristotelism to Christian learnings in 13th century, all were also alchemists. Roger Bacon is also known under the name **Doctor Mirabilis**.

Philosophical-alchemical doctrine has opened new views on the problem of subject, object, and the substance. As we have discussed in previous lectures, in chemistry, which have, in contrast to physics, developed from the alchemical and protochemical tradition of experimental approach rather than from the accurate analysis of the space-time, the *separation on the explorer and explored object* has been disrupted. The substance, and the manipulation with the substance, i.e. the alchemical practice, were not separated. As a combination of chemical operations known as **Opus Magnum** [5] was within the Middle Age chemistry intertwined with the personal spiritual experience of the experimenter. It would be very difficult to consider the alchemist as a subject, and the substance as an object. One of the most prominent alchemists **Basilius Valentinus** describes, nearly in the year 1600, **vitriol**, the blue stone (copper(II)-sulphate hexahydrate) as: *Visita interiora terrae rectificando invenies occultum lapidem* (in the free translation: Visit the interior of Earth, where you will, by your personal purification, create the stone of wisdom) . This description and principle was in the base of alchemical thinking. Here we can recognize the triad of alchemical philosophy: vitriol as substance-object, subjective auto-perfection of the alchemist, but also the acronym v i t r i o l. In the performance of alchemical experiments, the spiritual experience and sensation of the experimenter was not separated from the act of experiment. Thus, the notion about the unity of experimented and experimenter, the object and the subject, is originally alchemical.

Until the appearance of quantum mechanics and its Copenhagen interpretation, in the classical mechanistic Cartesian dualism, subject and object were clearly separated. The world was considered exclusively as an objective reality. The revolutionary change in thinking has appeared after the discovery of photons and electrons, which were interpreted as to have a dual nature, they can behave as particles and as waves. It has to be mentioned that the idea about the dual nature has caused a distrustfulness among the scientists, but also, on the other hand, this concept has been widely abused and welcomed in pseudo-sciences. However, the duality is not a sort of exotic property of particles, but rather the special relationship between the explorer and the explored. While the results of one experiment could be interpreted by the wave nature, for the instance of electron, other experiments were better explained by the behaviour of particles. Since the interpretation of the nature became dependent on the subjective choice of the experimenter, the intellectual base for the scientific revolution called quantum mechanics has been created. The untouched and independent nature explained as the superposition of all the possible states is for us unreachable and unknowable. The act of cognition is possible only after grasping of the curious man in this coherent situation, but on the way, that this grasp in the same time changes the observed nature in accordance with the intentions of the explorer. The reality is such as we wish to see it: the entanglement of the subject, observer, and the object, observed. Obviously, the duality of nature, as well as the destruction of the *subject-object* is omnipresent in scientific epistemology, but its ancient roots have its origin in protochemistry and alchemy.

Subjective-objective approach in alchemy is manifested also as the duality of *esoterism* and *exsoterism*, alchemy is speculative and operative. This duality is a source of the future scientific methodology, which consists from *theory* and *experiment*. However, alchemical speculative proto-theory becomes more and more introspective, and it has departed from the rational experimental practice. **Philipus Aureolus Theophrastus Bombastus ex Familia von Hohenheim Paracelsus** (1493–1541), the historical synonym for the European alchemist, has developed the esotery till the unexpected proportions. His "arcanic", secret language in alchemical scripts seems to be understandable only to himself. Paracelsian learning by which the disease is the consequence of the disturbation of the equilibrium between the three principles, salt, sulphur, and mercury, should be perceived as a metaphor intended for use only of the most lettered people. In spite to this, Paracelsus paves new ways that finally arrived to chemistry as a science. Although Paracelsus did not renounce the concept of four elements and three principles from which he has developed his alchemical theory of medicine, the *iatrochemistry*, he has transferred the idea about the reflection of events in haven to events on Earth into the relationship between microcosm and macrocosm. For Paracelsus, the observable world could be explainable only after our knowledge of the invisible microcosm. Thus, the visible reality is not only the hierarchical top-down relation from the superlunary immobile spheres, but also the bottom-up dynamics that origins from an elusive microcosm.

While the speculative alchemy circulated in vicious cycle of the word-play, the operative alchemy has progressed. Number of experiments has accumulated; the laboratory experience has grown with filling the laboratory books by adding new observations. At the end of 16th century, the observable experimental alchemy was almost completely unexplainable within the traditional Aristotelian concepts. The idea about four elements became useless!

**Jan Baptist van Helmont** (1580–1644), although he together with his esoteric statements belonged to the Paracelsian circle, has revolutionized his operative approach by moving from qualitative to the quantitative experiment. By measuring the mass of tree of willow, which has grown in the known amount of soil, van Helmont has shown that the mass of tree increases in some correlation with the amount of water used for sealing, keeping the amount of soil constant. Consequently, Earth cannot be an element. For van Helmont, fire and air are also not elements. Only water remained! Is it the return to Thales and his "proto-water", the return for more than two thousand years? All is known from ever, is not? By introducing measurement in manipulation with the substance van Helmont became a member of the intellectual circle of his generation, the Galilean circle, the society that was an embryo of the scientific community.

## 4.1 Conclusions

The practical, experimental approach to philosophy developed during the Hellenistic period in Alexandria has been transferred to Middle East where the new ideas were incorporated. To the Empedocles' four elements, earth, water, air, and fire, the three new principles (salt, mercury, and sulfur) were added. The practice of proto-chemistry was renamed to "alchemy", and its original philosophical intentions were extended with the new scope: its use in medicine. Alchemists hoped to prepare the substances which can provide us the eternal life. Experiments of alchemists were also combined with the personal mystical experiences of the alchemist. In such form alchemy moved to Europe where, in the time of scientific revolution, it has been transformed to chemistry.

## References

1. Grdenić D (2001) Povijest kemije, Liber, Zagreb (History of Chemistry, original in Croatian)
2. Brock W (1992) The fontana history of chemistry. Fontana Press, London, p 1992
3. Eco U (2016) I limiti dell'interpretazione. La nave di Teseo, Milano
4. Ficino M (1999) Scritti sull'astrologia. RCS Libri, S.p.A, Milano
5. Rabinovič VL (1979) Alhimija kak fenomen srednjvekovoj kulturi. Nauka, Moskva

# Occult Enlightenment

# 5

*The Divine Light is always in man, presenting itself to the senses and to the comprehension, but man rejects it.*

Giordano Bruno

**Abstract**

The appearance of scientific revolution in the period of Enlightenment is explained as a consequence of the interplay between three factors: mechanical philosophy, new approach in scientific education, and internationalization-"objectification" of science. I also argue how the relationship between scientific practice and public education is bidirectional: science must be in the base of education, but, conversely, the educational methods provide special requirements to science to be organized in a specific way. In addition, it is explained how determinism has appeared in science and philosophy.

**Keywords**

Enlightenment · Mechanical philosophy · Laplace's demon · Incommensurability paradox · Determinism · Educational revolution · Chemical affinity · Classification methods · Scientific method · Empiricism

Scientific, and especially pseudo-scientific discussion often fall in the *paradox of incommensurability* what generally means the explanation of the appearances in one area by conceptual system from another discipline. Galilean, or, if you wish Copernican confrontation whit the Church hierarchy was in principle the example of the paradox of incommensurability. Scientific interpretation of the new science was evaluated by starting from other field out of science, the field of theology, or better to say pseudo-theology. In contrast, the same paradox is actual also today: the



H. Venčik, *Philosophy of Chemistry*, Integrated Science 2,
https://doi.org/10.1007/978-3-030-69224-7_5

theologian concepts are evaluated by starting from the axiomatic of a rational science. However, I wish to argue that, in spite to its unacceptability, incommensurability has played, and still plays today, a stimulating force in development of new scientific and artistic thought. New ideas often are developing within the unusual circumstances. *Credo quia absurdum est!* was attributed to **Tertullianus** (**Quintus Septimius Florens Tertullianus**, cca 160–cca 240) who evidently suggested the problem of incommensurability in his specific way.

In 16th, and at the beginning of 17th century, three actual viewpoints were incommensurable, church-theologian, occult, and rational. The last two, represented by alchemy and mechanical philosophy were the origins of scientific revolution.

Occultism has in Europe two sources, the philosophy of Renaissance, and alchemy. **Cosimo il Vecchio** (1389–1464) the regnant of Florence has housed the Church Consilium in 1439 which had the scope to unify western, catholic, and eastern, the orthodox church. Besides the church fathers, the participants of the Consillium were also the most famous thinkers and philosophers of that time. One of them, **Giorgio Gemisto Pletone** (1355–1452) has proposed the general unification of all the religions, which were most influential on the development of the European culture, with the special accent on the occult Egyptian tradition and the pagane beliefs from ancient Greek and ancient Persia with its Zoroastrian religion. The personification of such unification Pletone has found in Platone and the subsequent Platonism. The occult component has been additionally supported by the discovery of the book *Hieroglyphica*, attributed (without arguments) to **Horapollo**. The book consists of interpretations of Egyptian hieroglyphic scripts as a symbolic language about the secret occult knowledge. In the same time, in Florence has circulated also the version of Trismegistos *Tabula Smaragdina*. Such mixture of occultism, and revitalization of the classical antiquity has triggered the appearance of new viewpoints, development of science and professional art. The ideal about the *Prisca theologia* in principle originate from the same Pletone. Marsilio Ficino has accepted these ideas, and, by suggestion of Cosimo il Vecchio, he has organized the Neoplatonic institutions. However, most of the participants of the Consillium did not support Pletone.

As we know from previous lectures, another source of occultism was alchemy. This occult discipline was heretic, and unacceptable in the countries which were under the strong influence of the Roman Curia (*Curia Romana*). Consequently, the most of alchemists were moved to protestant countries and, partially, in France where the Church was less sensitive on the Copernican changes of scientific paradigm. In these countries, new synthesis in natural philosophy has appeared: **Cartesius**, **René Descartes** (1596–1650) has finally unified geometry and arithmetic by introducing the numerical description and mathematical representations of geometrical objects. In natural sciences, the prevailing method became the quantitative approach.

Transformations of geometrical objects were in the base of the representation of the world considered as an un-ended and perfect mechanical machine. "Numerization" of geometry has encouraged the scientists to believe in the possibility of the precise calculations of all the astronomical appearances in the far future. **Pierre-Simon markiz d'Laplace** (1749–1827) has been obsessed by the discussion about the idea about the perfect cosmos-machine. In scientific history, this concept is well known under the name *Laplace's demon*: if you sre siting on a distant planet, and if you know all the starting conditions of the position and movement, all the future development is precisely determined! Such a way, *Determinism* became the basic scientific concept.

The possibility of quantitative representation of the world imagined as a machine requires that the reason, *ratio*, must be in the focus of scientific and philosophical activity. The worship of the reason as a new divinity has stimulated the appearance of new scientific paradigm, and the scientific revolution. New world-views were already on the horizon.

Together with various historical and social aspects, which will not be discussed because they go beyond this book and its author, it could be argued that the scientific revolution of 17th and 18th century has been consequence of interweaving of three important factors: *mechanical philosophy, democratization of education within the frame of rationalism, and objectification of science*.

Undoubtedly, mechanical philosophy is rooted from physical investigations, because the mechanisms are in principle the space-time systems that are measurable and calculable. In 18th century, such mechanicistic studies have grasped also some "softer" disciplines such as medicine. The concept of *human body–machine* appeared mostly in the systematic medicinal studies of Dutch scientist **Hermann Boerhaave** (1668–1738). It is interesting to mention that Boerhave, together with many of medics in that time, gave an important contribution to chemistry. Boerhaave, although he belonged to the iatrochemical concept of the Parecelsian circle, has reaching to new quantitative approach of mechanical philosophy. His sharp turn to quantitativity in chemistry is evident in his accurate measuring of masses and temperature. Some authors consider Boerhave as a predecessor of physical chemistry. However, as a chemist, Boerhave was suspended between mechanical-philosophical viewpoints and the occult and Aristotelian chemistry.

Eclecticism, the conjunction of incompatible, the occult with rational, was especially noticeable in the work of the establisher of classical mechanics, and one of the authors of new scientific paradigm **Isaac Newton** (1642–1717). In the scientific history, Newton has been systematized between physicists, but his alchemical activity was almost forgotten [1]. Mechanical movements of celestial bodies that follow the exact mathematically described laws have probably stimulated Newton to try to explain the movements in microcosmos on the similar way. Conscious, like Paracelsus, that the visible reality of substance is a consequence of an invisible micro organization, Newton has attended himself to alchemical experiments. Numerous laboratory books with precisely described alchemical experiments has remained after Newton who intended to understand the possible structure of the micro world that is analogous to the celestial mechanics. Clearly, Newton was in his science close to the concept of the discrete atomistic picture of matter. But yet, not in vain. Although his ambition to find mechanical laws in alchemical experiments has exceeded the real scientific practice of that time, Newton has still contributed to the developement the new important chemical concept: *the chemical affinity*. Newton has proposed two sorts of affinity. First is the *mechanical affinity* by which the elemental particles—say, atoms—attract each other on larger distance, and repel each other on smaller distances, and the second is the *elective affinity*, i.e. the inherent property of substances to react with other substances. In modern words, the elective affinity is a sort of chemical reactivity, the property of substances to react with each other with various rates. Introduction of this term was revolutionary for chemistry, because besides of the *classification of substances*, Newton has initiated also the *classification of reactivity*. Let us mention, that in the second half of the last century it has appeared a textbook of organic chemistry that was based on the classification of reactions, not on classifications of compounds! [2]. In addition, the Newtonian mechanical affinity opened the way to the future appearance of the most important chemical concept—*the chemical bond*!

Mechanical affinity was also in focus of interest of the Croatian scientist, the Jesuit **Rudjer Josip Bošković** (1711–1787). The relationship among the attractive and repulsive forces between the material points (atoms) Bošklvić describe analytically by the curve that these forces connect with the distance between material points [3].

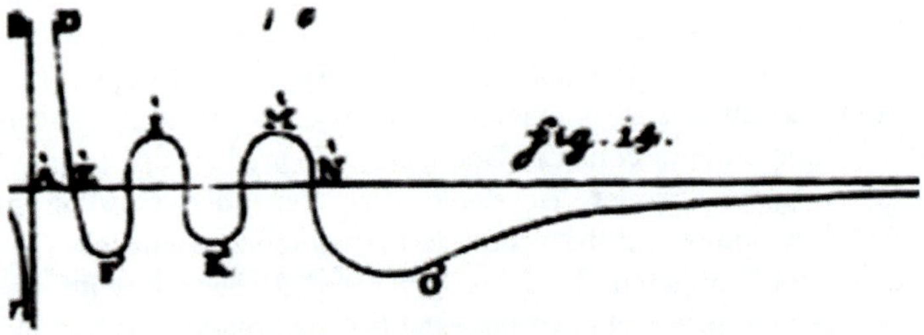


**Bošković curve that represents the dependence of attractive/repulsive force (ordinate) on the distance of material points (abscise).**

If we replace the force in the Bošković curve with the potential energy, we get the modern representation of chemical bond as a relation between potential energy and interatomic distance (*Morse curve*). As we know, the minimum of potential energy corresponds to the length of chemical bond.

**Morse curve that represents the dependence of potential energy ($E_p$) on the distance between atoms of chlorine and hydrogen in the molecule HCl ($r_{HCl}$)**

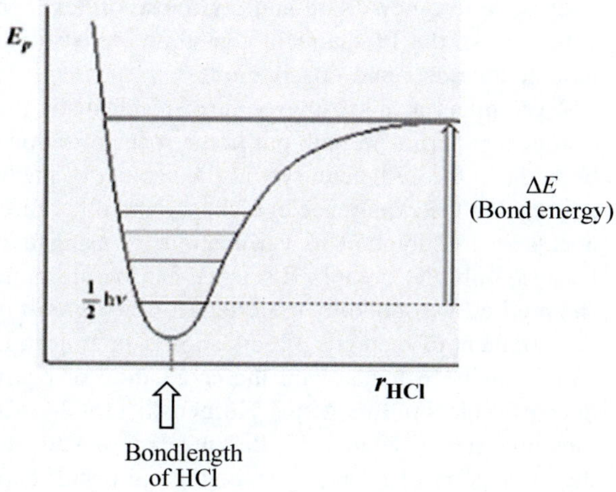


At first sight, both, Bošković and Morse curves are different because Bošković provides more than one minimum, i.e. more stable equilibrium distances between atoms. In should be mentioned that the foresight of Bošković was confirmed by modern investigations that have shown the reality of more stable minima between atoms, [4] the chemical bond, the distance that appears in crystals under specific conditions, and the very well-known **van der Waals** distance (**Johannes Diderik van der Waals**, 1837–1923) between the atoms that are not connected by chemical bond.

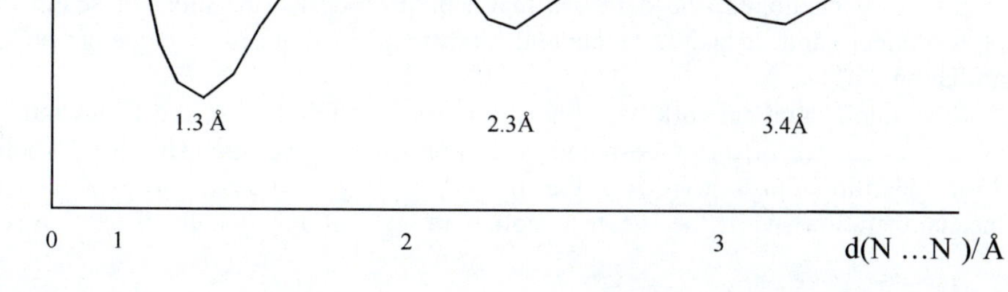


**Different equilibrium distances (minima on the potential energy curve) of nitrogen atoms in molecule. The analogy with the Bošković' ideas [4].**

In spite to a new light, and a glimpse of new concepts, alchemical experiments described in the literature of that time were still an almost arbitrary mixture of innumerable data and observations.

Experiment as a starting point of mechanistic thinking, and the traditional (al)-chemical experiment, both put in the focus of rational reasoning *the method*. I t can be said that the 17th century was the century of method. Since all earlier knowledge collected in the experience of alchemy and other disciplines was chaotic, it has been necessary to submit it to verification by using the rigorous systematic approach because only the organized science can be presented to community. **Focusing on the method** was not only important for the rigor of original science, but also for the development of methods of education. Copernican turn in the system of education, which has been focused on the acquisition of equilibrated knowledge, was introduced by bishop **Jan Amoš Komenski** (1592–1670), the Czech philosopher and methodicist. For Komenski, the integrity of knowledge, *the pansophia*, comprises the unification of religious, rational, and occult experience, as well as the way of thinking [5]. Isn't the spiritual alchemy exactly the discipline, which has for centuries been immersed in pansophian unification?

The utopian idea about the perfect free society, which got its climax in the Enlightenment, was realizable only through the new system of general education. I wish to argue that, as the scientific methodology has caused the change in methodics of education, the same education made a pressure on science to organize itself on a new way. Systematic evaluation of the present chaotic scientific experiences has triggered new scientific knowledge, and initiation of new intellectual paradigm.

It has been necessary to organize also the alchemical knowledge by separating its important from its unimportant content. This was the beginning of the creation of the chemical conceptual system. The result was the appearance of the first alchemical textbook. Medic and alchemist **Andreas Libavius** (1555–1616) has published the alchemical textbook *Alchemia*, on the threshold of the 17th century, in the year 1597.

To be efficient, the democratization of education should involve in itself the objectification and the internationalization of the knowledge by establishing corresponding standards and conventions. Step-out from the frames of the closed impenetrable circles of scientists is possible only by the establishment of a sort of international catalogue, the database that would comprise the most of scientists, philosophers, and ideas, independently on their institutions and geographical affiliations.

Such intellectual network has appeared during the first half of the 17th century, as a consequence of the widespread correspondence of **Samuel Hartlib** (?1600–1662). Hartlib's epistolary circle, the *Invisible* College, [6] started to expand with the cooperation with the nearest collaborators, among which were already

**Facsimile of the cover of the first textbook of alchemy written by Andreas Libavius from 1597**

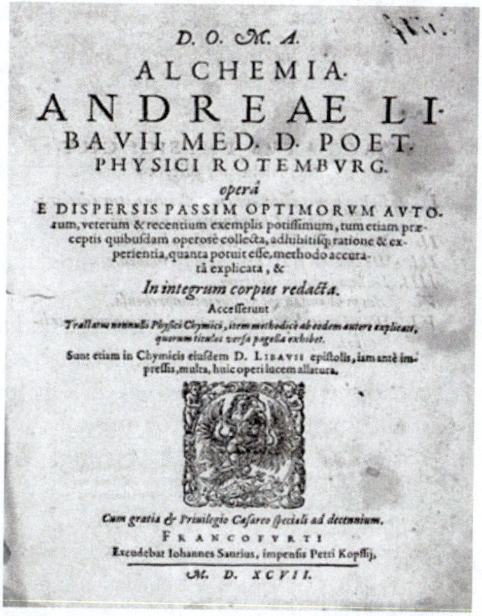
D. O. M. A.
ALCHEMIA
ANDREAE LI-
BAVII MED. D. POET.
PHYSICI ROTEMBURG.
operâ
E DISPERSIS PASSIM OPTIMORUM AUTO-
rum, veterum & recentium exemplis potissimum, tum etiam præ-
ceptis quibusdam operosè collecta, adhibitisq; ratione & ex-
perientia, quanta potuit esse, methodo accura-
tâ explicata, &
*In integrum corpus redacta.*
Accesserunt
*Tractatus nonnulli Physici Chymici, item methodicè ab eodem autore explicati, quorum titulos versa pagella exhibet.*
Sunt etiam in Chymicis eiusdem D. LIBAVII epistolis, iam antè impressis, multa, huic operi lucem allatura.
*Cum gratia & Privilegio Cæsareo speciali ad decennium.*
FRANCOFURTI
Excudebat Iohannes Saurius, impensis Petri Kopffij.
M. D. XCVII.

mentioned Ianus Commenius, but also the reformer of alchemy **Robert Boyle** (1627–1691). The database that comprised not only scientists, but also other practitioners in the field of humanities and economy, all from the civilised circle of that time, has started to grow. Hartlib has stimulated collaboration between scientists by joining them and informing them about actual topics of research, opinions, and discoveries. The established network has been transformed by gradual formation of thc principles of scientific universality.

This social objectification has mirrored in already mentioned methodological objectification. Scientific true was not more a matter of individuals, or a matter of particular closed circles, but rather it was oppened to the community. In such a way, *criticism* and openness in scientific activity begun to play the key role; science became step by step non-personal. This social influence on the science, as well as already mentioned universalization of education have as a result the objectification of science, and, finally it led to the scientific revolution.

The focus of scientific investigations has been shifted to a quantitative experiment. Traditional alchemy of Empedoclean principles did not more match in this new scientific-methodological scheme. Within these frames the idea element-principle was out of sense.

As we have mentioned, van Helmont has clearly rejected the concept about four elements. For him, only water is acceptable as a sort of elementary substance. All things, those measurable and variable are only a specific form of water. Water is not more a sort of Thalesian element.

The term element makes sense only as a real empirical and mechanical substance that we can measure, and with which we can manipulate! *Empiricism* as an approach to the world is booming. All the knowledge has its origin in the experience, and it is measurable and law-like. This must become the object of the general education of the widest society.

## 5.1 Conclusions

Scientific, social and intellectual movement known as Enlightenment has appeared as a consequence of the combination of mechanical philosophy, development of public education, and objectification and internationalization of science. It is argued how the educational methodic forced the science to be organized. First textbooks of chemistry which appeared in this period have an important consequence: the introduction of classification methods in chemistry and in science.

## References

1. Dobbs BJT (1975) The foundations of Newton's Alchemy. Cambridge University Press, Cambridge
2. Hendrickson JB, Cram DJ, Hammond GS (1970) Organic Chemistry, International edn. McGrow-Hill Kogakusha, Tokyo
3. Bošković RJ (1763) Theoria philosophiae naturalis, Typographia Remondiniana, Croatian Edition 1974. Liber, Zagreb
4. Varga K, Volarić J, Vančik H (2015) Crystal disordering and organic solid-state reactions. CrystEngComm 17:1434
5. Comenius JA (1960) De rerum humanarum emendatione consultatio catholica—deutsch: Universeller Ratschlag für die Besserung der menschlichen Dinge – 1642–1670; lat./dt. Heidelberg (u: Pampaedia, Hersg. D. Cyzevs'kyj und H. Geißler)
6. Webster Ch (ed) (1970) Samuel Hartlib and the Advancement of Learning. Cambridge University Press, Cambridge

# 6 Conceptualization of Science and Experimental Model Systems

*The aim of science is to seek the simplest explanations of complex facts. We are apt to fall into the error of thinking that the facts are simple because simplicity is the goal of our quest.*

A. N. Whitehead

**Abstract**

This chapter is focused on some aspects of scientific methodology. The importance of the choice of the investigating models, and how this model selection is critical for the formation of the new scientific paradigm is discussed. I demonstrate how the selection of combustion process as a model for the investigation of chemical transformtions led to the formation of the first chemical theory—the phlogiston theory.

**Keywords**

Models in science · Phlogiston theory · *Ad hoc* theory

Although scientific concepts are focused on the particular topics, their epistemological structure must be universal. History of science is full of examples, which show that scientific revolutions, the changes of paradigm, were started with the investigations of very specific case studies, rather than with the research of great very well known systems.

The transition from classical to quantum physics did not in principle occurred within the actual system of Maxwell (**James Clerk Maxwell**, 1831–1879) electrodynamics that represented a triumph of classical physics at the end of 19th century. Quantum theory appeared from in that time almost irrelevant observations of radiation of a black body. **Max Planck** (1858–1947) was not spurred by the



H. Vančik, *Philosophy of Chemistry*, Integrated Science 2,
https://doi.org/10.1007/978-3-030-69224-7_6

scientific trend to be focused on the main-stream problematics. Rather, he was guided by his sincere scientific curiosity. In spite to that, the laws of the irradiation of black body were among the cornerstones of the paradigmatic change of the entire physical concept. The quite narrow specialized study has opened the ways to new universal principles. New science is hidden in details!

What is the secret of Planck's method? Max Planck has found the new *experimental model system*, the collection of in this time neglected appearances, whose nature is such that they can unveil new lawfulness of the universal importance. Many, although not all, the "accidental discoveries" could be classified in this category.

The transition from the occult alchemy, in which the irreproducible experiments have prevailed, to the systematical rigorous science was also a consequence of focusing of chemists to a specific chemical phenomenon—the combustion. As an experimental model system, combustion possesses particular characteristics, which enabled creation of universal laws, and, consequently, the appearance of the first chemical scientific concept.

**Georg Ernst Stahl** (1659–1734), professor of medicine on the German university Halle has represented the first rational chemical theory starting from the systematic studies of the combustion reaction [1, 2]. The combustion of metal yields a substance that has been known from the time of alchemy as calx (today we know that it is a metal oxide). Stahl has proposed a concept by which during the combustion the metal releases a subtle substance called *phlogiston* (φλογιστόν, combustion, φλόξ, flame) [3]. For instance, metal is by combustion transformed to calx and phlogiston:

$$\text{METAL} \rightarrow \text{CALX} + \text{PHLOGISTON}$$

Since every theory must be internally coherent and not contradictory, this phlogiston theory has to answer the question about the fate of phlogiston after combustion, and whether it is possible to add phlogiston back to calx, and to recover the original metal. By Stahl, there are substances, for instance coal, which are rich of phlogiston. If this is true, then heating the mixture of calx and coal should yield metal. This fact is known already from alchemical practice. In this way phlogiston theory becomes universal; it proposes the *existence of an elusive stuff that could be exchanged between substances.*

$$\text{METAL} \rightarrow \text{CALX} + \text{PHLOGISTON}$$
$$\text{CALX} + \text{PHLOGISTON (from coal)} \rightarrow \text{METAL}$$

It is important to mention that Stahl allows the possibility of the existence of a stuff that is *unobservable by our natural senses.* Thus, phlogiston is a sort of a *conceptual substance.* So, but what were atoms in antiquity but the conceptual particles! As a rule, any microstructure was conceptual before its rational perception.

Phlogiston theory has been the first chemical theory, which appeared from of the investigations of the specific chemical process, but it became universal because it describes not only the actual phenomena but also predicts the nature of the future processes that will be experimentally studied [4]. In addition, phlogiston theory had also unifying power, because it classifies combustion, breathing and oxido-reduction processes in the same category.

However, phlogiston theory has one important deficiency: it has been only qualitative, there was no mathematical apparatus accessible for it. This weakness has been crucial for its viability. In spite to that, phlogiston theory has initiated the formation of another chemical theory, which, starting with more rigorous definition of chemical elements, became the foundation of modern chemistry.

The descent of phlogiston theory (tentatively speaking) is an example of the Popperian approach by which a good theory must be falsifiable, rather than verifiable. From Popperian view a good theory should contain the inherent possibility of testing. There are two of such tests in the phlogiston theory, the first is the quantitative verification on the combustion reactions, and the second is qualitative.

The first test that rejects this theory is quite evident, the mass of the combusted metal (i.e. the mass of calx) must be smaller than the mass of the reactant-metal because the metal has lost phlogiston by combustion. However, this contradicts observations since the mass of calx (the metal oxide) is always greater than the mass of metal before combustion. The theories, which did not passed tests were not completely rejected. Instead, they were simply extended by adding so called *ad hoc theories*, the concepts that cannot be independently proven. Such *ad hoc* theory invented to "save" the phlogiston theory affords that phlogiston possesses a **negative mass**!

Qualitative test that refutes phlogiston theory is based on the observation that it is possible to transform calx to metal without adding phlogiston (for instance by heating with coal). **Pierre Bayen** (1725–1798) has carried out such experiment in 1774: by heating the red mercury oxide (in that time called, *red stone*, of *mercury calx*) without the presence of coal as a donor of phlogiston he prepared the starting metal—mercury. Bayen has also observed the evolution of a gas, besides the appearance of mercury, and that this gas can, in the reaction with mercury, yields calx again. However, Bayen did not perceive that he has really discovered a new chemical element, oxygen, and that this experiment is on a threshold of scientific revolution. This paradigmatic move belonged to **Antoine Laurent Lavoisier** (1743–1794).

This phenomenon of combustion as a "model system" in science, which has initiated the change of paradigm in chemistry, was in the experiment with red stone extended with a new "submodels system", the investigation of the nature of air. Besides combustion, there are also other observed chemical transformations in which air is immediately included. Even phlogistonists were noted that air is not the same before and after combustion of a substance: the air before combustion is without phlogiston (*dephlogisticated air*), but after combustion the air is changed because it is full of phlogiston (*phlogisticated air*):

$$\text{METAL} + \text{DEPHLOGISTICATED AIR} \rightarrow \text{CALX} + \text{PHLOGISTICATED AIR}$$

If something is glowing in the closed space, the surrounding air receives also phlogiston as addition, and it would be expected the increase of volume. This phlogistonian prediction also did not passed the Popperian test of the refutability of theory: in all the experiments with combustion in the closed container, the observations were opposite, the phlogisticated air had smaller volume than dephlogisticated air, namely for one fifth!

It is important to point out, that, in spite to the popularity of the phlogiston theory, some scientists were occasionally more or less sceptical. **Pierre-Joseph Macquer** (1718–1784) is, in his at that time very influential dictionary of chemistry, in doubt regarding the definition of phlogiston. Sometimes phlogiston is a combination of air and fire, but sometimes phlogiston is only a sort of chemical "method".

The model system "combustion – air", which played the crucial role in the formation of new science, has not been chosen consciously by particular scientist. Rather, it was the result of the particular circumstances in the historical development of scientific research. If this is true, than paradigmatic changes are mere a sort of self-organization, a crystallization of new knowledges. **It is not possible to manage the development of science!**

Phlogiston theory has also an additional disadvantage: it was not applicable for the combustion of organic compounds, the substances from plant and animal origin. Was this already the beginning of *vitalist* thought about the two different chemistries, chemistry of life, and chemistry of inorganic world?

## 6.1 Conclusions

By analyzing the appearance of the first scientific theory in chemistry, the phlogiston theory, I argue that focusing on the proper model system is crucial for paradigmatic progress. For instance, for the discovery of quantum theory such system was the black body radiation. For scientific revolution in chemistry, the model system was the reaction of combustion.

## References

1. Stahl GE (1715, 1740) Opusculum chymico-physico-medicum, Halle et Magdeburgiae
2. Kirwan R (1789) An essay on phlogiston and the constitution of acids, Reprint in 1968, Frank Cass, London
3. Paušek-Baždar S (1994) Flogistonska teorija u Hrvata. HAZU, Zagreb (In Croatian)
4. Chang H (2010) The hidden history of phlogiston. HYLE 16:47

# Ceteris Paribus

# 7

*We think in generalities, but we live in detail.*

A. N. Whitehead


## Abstract

All the scientific theories, laws and concepts are valid only within the a priori given conditions and circumstances: the ceteris paribus principle. Here I argue how from this principle it follows that we cannot talk about the natural laws but rather about the laws of particular sciences.



## Keywords

Natural laws · Scientific laws · Ceteris paribus principle · Air as a model · Gas as a model · Element · Elementary stuff · Method of correlation


**Henry Cavendish** (1731–1810), an English chemist, who was student in Cambridge, the University that he abandoned before finishing his study, has performed experiments in his private laboratory. In the work *Experiments on factitious airs* (1766), Cavendish is discussing about the air that could be fixed in other bodies: the *fixed air*. Cavendish has argued that there are various sorts of airs which could be fixed, for instance the *flammable air* (hydrogen), or *fixed (bound) air* (carbon dioxide). Thus, all the airs could be free or fixed in other bodies. Flammable air is fixed in metals (zinc or iron) from which Cavendish can release it by the reaction with sulphuric acid. It is considered in the history of chemistry, that Cavendish work is the most important contribution in the discovery of hydrogen.

Fixed air was a point of interest also in the studies of other chemists, especially of **Joseph Black** (1728–1799): the air that is releasing from the carbonate salts in the reaction with, for instance, sulphuric acid. Similar sort of air has observed




H. Vančik, *Philosophy of Chemistry*, Integrated Science 2,
https://doi.org/10.1007/978-3-030-69224-7_7

**Stephen Hales** (1677–1771) who has isolated it as a product of fermentation, and degradation of biological material.

The air that could be fixed or released, used as a scientific model system, has played a critical role not only in the final rejection of the idea about four elements, but also in the introduction of the discrete—the atomistic concept about the structure of matter. The scientific sceptic **Robert Boyle** (1627–1691) in his work *The Sceptical Chymist* (1661), written as a dramatic dialogue among the four representatives of different opinions, has finally discarded the concept about the four elements that has been useless in explanations of a plenty of observations collected during the long history of alchemical practice. Boyle has developed the idea about elements in two directions, from the opinion that this concept is unnecessary at all, till the redefinition of the element as a "materialized category", i.e. directly related to the experimental practice. The notion "element" is in such a way transformed into "elementary stuff" that can be characterized by observations, for instance as a substance that cannot be decomposed by chemical methods. Such a transition from the metaphysical to experienced elementary matter [1] is the excellent example of the empiricistic epistemology that was actual in the period of Enlightenment. It is important to point out that from such a "materialistic" definition of element follows the opinion, that the number of elements is not closed. Thus, the World, the Universe is open!

After these Boylean examinations, the discussion about elements has started its road towards the scientific revolution in chemistry: what is chemical element at all, the substance, or an abstract concept? Boyle stood up also for the revitalization of the atomistic representation of microcosm. *Minima naturalia*, how Boyle has called the Leucippus-Democritian atoms are the basic particles of elements, the barriers of their observable properties.

Robert Boyle has his interest for airs (gases) extended with the additional examinations in the way, that this new methodology became one of the key factors which have played the role in the appearance of scientific revolution. He has examined the relationship between pressure and volume of air by introducing accurate measuring and the *method of correlation* of the obtained parameters. The dependence of pressure on the volume is possible to analyse quantitatively only in one case: if all the other conditions are unchanged—constant—the principle which is in epistemology called *Ceteris paribus* [2]. In the Boyles example, the temperature should be constant, and the experimental system should be isolated. Experiment *in isolated systems* became the basic condition for all the quantitative approaches to research. The result was the rigorous law about the constant product of pressure and volume –the Boyle-Mariotte law. **Edme Mariotte** (1620–1684) has independently discovered the same law.

*Ceteris paribus* methodology has opened the door for new science. Remember that all physical laws are *Ceteris paribus* laws, what means that they are valid only within the a priori given conditions and presumptions [3].

Boyles contemporary **John Mayow** (1641–1679) has systematically built the isolated experimental constructions for the study of various airs. His experiment with the candle that float on water and glows below the glass cylinder is known from the elemental school textbooks. This experiment in the isolated system is far-reaching. When the candle is burnt, the water level ascends for one fifth of the volume. Mayow has performed a series of similar experiments in the isolated systems, i.e. under the *Ceteris paribus* conditions. In such a way Mayow made the chemical experiment quantitative. He has accurately measured the volumes and masses of the reactants (the substances whith which begins chemical reaction) and products. It can be recognized from the documents, that Mayow had *de facto* already discovered that the air consists from two different components, of which one participates in combustion by fixing itself in the burnt substances. The final discovery of *oxygen* and its role in combustion was already forthcoming.

Quantification of the experiment has been the step by which chemistry, although it was traditionally established on the qualitative observations, became mathematized by using the fundamental principles of physical methodology. Chemistry and physics begun to be overlapped in methodology. However, this fact does not mean that chemistry and physics were unified in their epistemology, i.e. that new chemical laws and concepts became reducible on, or deducible from physical laws. Newton's alchemical experiments have been an unsuccessful attempt to unify these two epistemologies, as it was discussed in fifth lecture of this book.

## 7.1 Conclusions

Starting from the Robert Boyle's law about the relation between the volume and pressure of gases I argue that the scientific laws and concepts are possible only within the model system with a priori defined conditions. In Boyle's example, all the parameters except the volume and pressure should be fixed. This principle, the *ceteris paribus* principle, is one of the basic tools in scientific epistemology.

## References

1. Boyle R (1772) The Works, T. Birch, ed. reprint (1965) Hildesheim
2. van Brakel J (2000) Scerri E, McIntyre L, (eds.) Philosophy of chemistry. Boston Studies in the philosophy and history of science, p 306
3. van Brakel J (2012) Chemical engineering science, p 540. In: Thagard P, Woody A, Findlay Hendry R, Woods J, Needham P (eds) Philosophy of science. Elsevier

# Scientific Revolution in Chemistry

# 8

*Reasoning is nothing more than a language well arranged.*

A. L. Lavoisier


**Abstract**

Scientific revolution in chemistry has appeared at the end of eighteenth century. This period of science is bound to the name of Lavoisier, who, by using proper experimental model, has developed the most important concept of chemistry: the definition of chemical element. The case study about the rejection of the phlogiston theory after the discovery of the element oxygen is dedicated to the general discussion in the philosophy of science about the extent and limitations of the radical falsificationism.



**Keywords**

Heurism · Falsificationism · Oxygen · Chemical element · Scientific revolution in chemistry · Method of classification · Method of systematization · Analogy · Transduction · Nomenclature · Taxonomy · Science and linguistics · Ockham razor · Substitution reaction


From the work of Thomas Kuhn, we know that the period of scientific revolution is characterized by changing the paradigm, for instance, the geocentric planetary system was irreversibly replaced with the heliocentric one. Newton has revolutionized physics by introducing new *heuristic* parameters: force and mass. The new light in science is appearing, the new "design" of the scientific modes of viewing.

Since chemistry origins from experiment, it was in the longest historical period the qualitative discipline. This fact is important from the philosophical point of






H. Vančik, *Philosophy of Chemistry*, Integrated Science 2,
https://doi.org/10.1007/978-3-030-69224-7_8

view, because in that way chemistry may to say something important about the *quality* as a metaphysical category. The manipulations with qualities requires a special methodology that is *based on analogies* rather than on logics. Within this epistemology, the main method that was developed is—*classification and systematization*. Similar, although not equivalent, development has appeared in the history of biology, and geology—mineralogy.

*Ceteris paribus* quantification of the chemical method has induced the extension of chemistry, which has been critical for the change of paradigm. Historically, many of scientific revolutions were related to the name of particular scientist. In chemistry, it is undoubtedly **Antoine-Laurent de Lavoisier**. First developed chemical theory, the phlogiston theory that was the excellent achievement of the qualitative science, the chosen stimulating model systems (experimentation with air, combustion, fixing of gases), and the rigorous quantification of experiments in the isolated conditions, laid the foundations of the scientific revolution.

What would be the content of new chemical paradigm? First of all, it is necessary to clarify the notion of element and to consider it within the traditional *method of classification*. The next step is to find the relation of element with the atomistic theory, as well as to open the ways for the future discovery of the structure of microcosm and its reflection on the macroscopic observation. Qualitative experiments and measurable data should be meaningfully associated to corresponding abstract models of the invisible structure—this is known as *the method of transduction*.

Lavoisier has initiated this avalanche of paradigmatic changes by discussing the concept of elements, and the possibility of their classification starting from the rigorous experimentations (Lavoisier, 1789).

As I tried to show, every important scientific step begins with the selection of the convenient model system. For Lavoisier, the model system was focused on the "purest part of the air that supports combustion, and, in reactions with non-metals, it yields acid—**the element oxygen**". On the ceremonially Easter lecture in the French Academy 1778 Lavoisier has emphasized:

> The principle that is unifying with metals, which than are calxified, and which increase their weight is: nothing else than the purest, and the healthiest part of air, which, after its combination with metals, can be free again.

It is commonly in the history of science, that new discovery is already in practical circulation, but that the scientists are not aware of it. Special persons like Lavoisier are necessary for the recognition of the almost evident discovery. The discovery never origins from a coincidence, but rather from the ability of **recognition**!

Oxygen with its main properties has been in principle already discovered by Scheele, Pristley, and, honestly, by Mayow. However, none of them did attribute it to elements, and, none of them was able to formulate new universal chemical principles from this discovery.

For instance, **Joseph Pristley** (1733–1804) has prepared the "dephlogisticated air", and described its unusual properties, alleviation of breathing, and supporting of combustion. Or, **Carl Wilhelm Scheele** (1742–1786) performs similar experiments by heating salts which have released "fire air" characterised by the same properties as the "dephlogisticated air" prepared by Pristley. Evidently, the obsession with Empedoclean-Aristotelian concept that air is an element was so strong, that, in spite to the clear experimental observations, it has "blocked" every glance that air could be a mixture of differing elemental substances.

Lavoisier had a sort of enlightenment and courage to see in this discovery more than an isolated phenomenon and the explanation of only one category of chemical changes. He steps out from the routine of the phlogiston theory with the idea about the air as an element, and he has changed the paradigm by showing that the "dephlogisticated air" is really the chemical element, i.e. mere it is one component of air. But, what about water?

Hydrogen, which Cavendish called "flammable air", was a subject of experiments of a large number of chemists during the 18th century. By systematic experimentation, Cavendish has shown that his "flammable air" explodes in the mixture with "dephlogisticated air", if this mixture is in ration 2,02 : 1, what he has precisely measured. The product of this explosion was water in the form of drops collected on the walls of the closed vessel in which the reaction has occurred. Just to remember, the "dephlogisticated air" is a common air in which metals and non-metals can burn, and "phlogisticated air" appears after the combustion when phlogiston is released from burning substances.

What would be more logical than the fact, that these two airs, flammable and dephlogistonized, are two different gaseous substances whose mixture explodes and yields water. But then, water is also not an element. Again, the inability to step out from the Empedoclean paradigm about water as an element has prevented the logical conclusion about water as a compound that consists from two gases, hydrogen (flammable air), and oxygen (dephlogisticated air). Instead, Pristley and Scheele have explained their experiments within the phlogiston concept: the flammable air is the phlogistonized water (water + phlogiston), and the dephlogistonized air is really the dephlogistonized water (water—phlogiston), thus,

$$(\text{WATER} + \text{PHLOGISTON}) + (\text{WATER} - \text{PHLOGISTON}) \rightarrow \text{WATER}$$

For phlogistonists water is still the element, which can fix phlogiston! One of the main characteristics of new paradigm is the introduction of new concept that simplifies explanation of observed appearances. Consequently, scientific revolution arise in the historical periods when the actual science starts to explain the phenomena with more and more complex and abstract representations. Basic property of the Copernican system is the simplification of interpretation of planetary movements. Remember, that the explanations of the planetary movements became extraordinary complicated within the geocentric system: introduction of the spiral (epicycle) orbits of planets, and the system proposed by Danish astronomer and alchemist **Tycho Brahe** (1546–1601) in which Venus and Mercury rotate around Sun, but all together rotate around Earth. Scientific revolutions in principle make simplifications, and the new paradigms are more general in relation to the old ones.

The simplicity as a criterion of the scientific thinking is known in the theory about science as the *Ockham razor* (**William of Ockham**, cca. 1287–1347). Willian of Ockham, the English Franchescan, scholastic and theologian, has revitalized the old idea already known between Pythagoreans, that among various explanations, the most acceptable is the simplest. Evidently, the step to simplification is the step in new epistemic world. However, Ockham razor as a methodological approach in the evaluation of science should be applied with care. This will be discussed in my last lectures.

Systematic experiments with water have continued and Lavoisier has performed many of them in collaboration with Laplace (**Pierre-Simon markiz d'Laplace**, 1749–1827). On the Day of St. Martin 1783, Lavoisier has, in the French Academy of Science, presented his discovery about water as a chemical compound that consists from gasses *hydrogen* ("the water-maker"), and *oxygen* ("the acid-maker"). In French language, the term *principe oxygene* means the *"acidifying principle"*.

The experiment, which Lavoisier has performed for arguing that water can be decomposed in hydrogen and oxygen, is far-reaching. The basic design of the experiment is the reaction of the hot water vapour with the red-hot chunks of iron. Oxygen from water binds to iron, and hydrogen is releasing. The appearance of hydrogen is proven by demonstrating its combustion at the exit nozzle of the apparatus. The importance of this experiment is not only in the fact that water is composed from oxygen and hydrogen, but also in the new approach in chemistry. *The substance*, in this case water, can by *a rational way be composed* from elements hydrogen and oxygen, as Pristley has demonstrated, and *decomposed in elements by the rational experimental procedure*, as Lavoisier has shown. Moreover, in Lavoisier's experiment is evident also another point: hydrogen that was bound to oxygen in water is *substituted* with iron—the *concept of the reaction mechanism* that will appear only in the 20th century chemistry is here announced. In this particular case, it is *the reaction of substitution*.

What is remaining, is the definition of the flammable air. Hydrogen and oxygen are chemical elements, but water is not. Lavoisier clearly shows that other combustion reactions are nothing else but the binding with oxygen. Burning of metals yields calx (oxides), or, phosphorus or sulphur combine with oxygen producing compounds, which in reaction with water led to acids. Thus, the dephlogistonizing air is the chemical element that yield acids—*oxygene* (ὀξύς *oxys*, acidic). The term "phlogiston" became needless, it has been removed by Ockham razor!

Oxygen is element, one of components of air. Water is chemical compound. What about the three thousand old idea about four Empedoclean elements? But, what are elements for Lavoisier?

The old-standing notions as earth, water and air have in Lavoisier's chemistry more universal meaning: they are *the aggregate states: solid, liquid, and gaseous*. Chemical element requires a new definition similar to that proposed by Robert Boyle. Newly defined chemical element as a substance that cannot be decomposed to simpler substances becomes the epistemological base of new scientific paradigm. By looking broader, this concept has initiated the journey towards the discovery of the structure of matter. In 1789 Lavoisier has published the *list and classification of chemical elements* in his book *Traité Élémentaire de Chimie*, the work that is the starting point in this journey [1]. The elements are classified in four groups, as it is shown in the following table.

First of all, it should be mentioned, that we know today that some of Lavoisier's elements, as, for instance heat and light, are not elements. The same is valid also for some acids. In addition, arsenic and antimony are within the modern chemistry not classified as metals. However, the importance of this classification is far-reaching from another reason. Namely, the division of elements in metals and non-metals is the beginning of the development of chemical theory towards one of the central chemical concepts—*the structural theory*. What is the scope of this classification? In following pictures it is demonstrated what is going on from the mutual combinations of the elements from the different Lavoisier's groups.

List and classification of chemical elements in the work Traité Élémentaire de Chimie (Lavoisier 1789)

| Elastic bodies | Light, heat, oxygen, nitrogen, hydrogen |
|---|---|
| Non-metals that can be oxidized and which form acids | Sulphur, phosphorus, carbon, hydrochloric acid, hydrofluoric acid, boric acid |
| Metals which can be oxidized and which can neutralize acids in salts | Antimony, arsenic, silver, bismuth, cobalt, copper, tin, iron, manganese, mercury, molybdenum, nickel, gold, platinum, lead, tungsten, zinc |
| Earths | Calxes (oxides) of magnesium, barium, aluminium, silicon |

In the first scheme bellow are shown the labels for the basic units from the theory of Lavoisier, the different metals ($M_1$, $M_2$, …), and non-metals ($N_1$, $N_2$,…), oxygen and water. It is demonstrated that, in the reaction with oxygen, metals forms calxs, the metal oxides.

The second and the third scheme are the essential shift towards the future structural theory. By combustion with oxygen, non-metals yield the non-metal oxides, which in turn, react with water to form acids. The third scheme shows that, depending on the choice of the metal or non-metal, various acids and their salts can be formed. It is important to point out that it is possible to transform one salt in another salt by *replacing either metal or non-metallic component*—they are exchanging by *substitution*. Metals or non-metals are by choice variable. However, as it is shown in the third scheme we have to indicate that the **salt as a category** is represented by the general graph (in the middle)—an *abstract representation of salt*, the concept which will evolve in the new chemical structural theory.

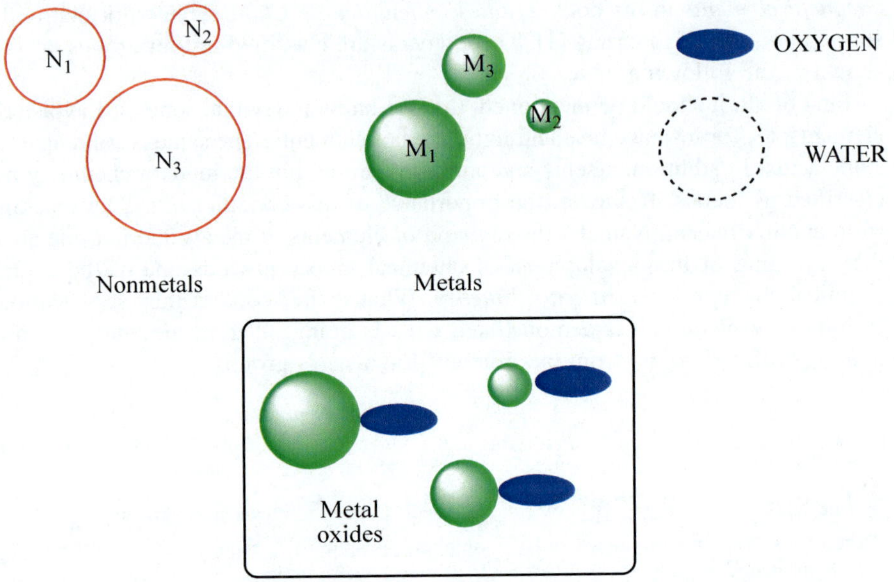


**Scheme 1**

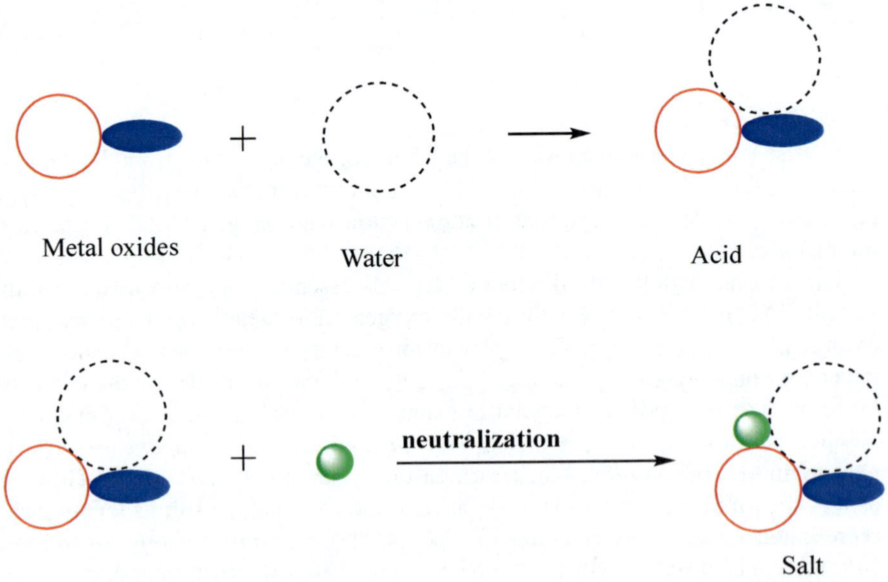


**Scheme 2**

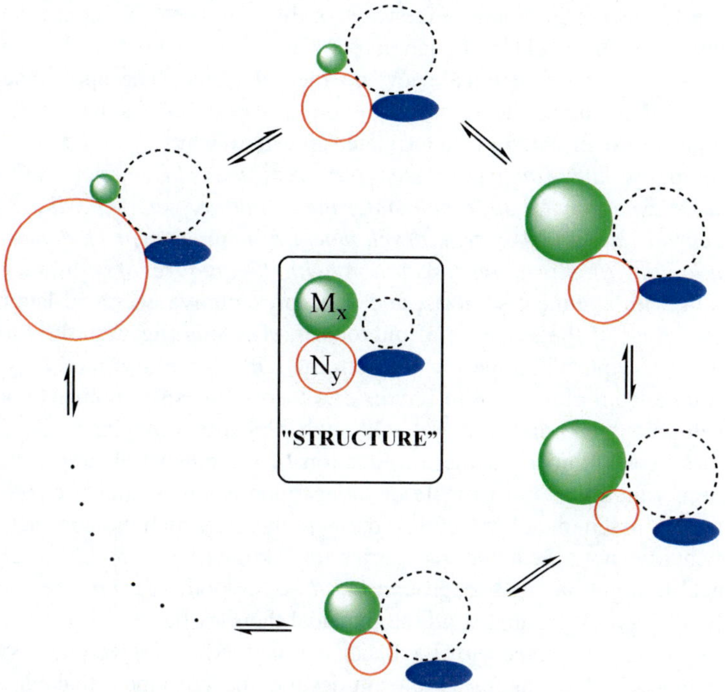


**Scheme 3**

The beginning of the structural theory has included the conversion of *chemical taxonomy* (the classification of substances) in *chemical nomenclature*, the syntax for the naming chemical compounds. The chemical name, formed by the rules of nomenclature, is the linguistic analogy to the chemical composition and microstructure of the corresponding substance. The components of the chemical name correspond to the components from which the compound is formed.

The first such nomenclature, which is the direct consequence of the theory of Lavoisier, was published in 1787. Besides Lavoisier, the authors were **Antoine François conte de Fourcroy** (1755–1809), **Louis-Bernard Guyton Baron de Morveau** 1737–1816), **Claude Louis Berthollet** (1748–1822). The discovery that calx could be prepared from metal and oxygen is visible in the name of this calx. For instance, binding of *magnesium* with *oxygen* yields calx, which should be called *magnesium oxide*. Thus, from the nomenclature name we can recognize (i) the composition of the compound, and (ii) *the way how it could be synthesized from elements*. Consequently, the scope of the systematic nomenclature is a method by which the *way of preparation of particular compound is codified in the chemical name*. The nomenclature name is not mere an iconic representation of a substance, but it is a bearer of the wide spectrum of information. For understanding the language of nomenclature the interpreter is unnecessary. Instead, mastering the nomenclature system requires the knowledge of the laws of science.

Evidently, the development of the ways of thinking is closely connected with the development of the structure of language. The secret alchemical language was a representation of the alchemical view on the substance. The appearance of new language is inherent in the scientific revolution. Symbols, substances, and phenomena are in novel paradigm interrelated in a new way.

In summary, *the Lavoisier's step out from the old science comprises the quantitative experimental approach under the conditions ceteris paribus, the choice of the appropriate model system, development of new concept of element, and the associating the linguistic representation with the nature of substance*. Experimentations with chemical subjects, and the representative chemical language have opened the door of the structure of microcosm. The missing step, the formation of the chemical script, will appear more than the half of century later.

The move from phlogiston to Lavoisier's theory has also another importance in the development of epistemology. Although elegantly imagined, the phlogiston theory has been founded on the introduction of the elusive abstract category, the phlogiston, which was not isolable or independently investigiable experimentally. As I have discussed in earlier lectures, the hypotheses, which include such imagined experimentally not verifiable categories are known as *ad hoc theories*. All alchemical thought has its background in *ad hoc* hypotheses. Let see the books of Paracelsus! Even more, almost all the physical theories before Copernicus were *ad hoc* theories. The new theory of Lavoisier is imagined exclusively as a system that mutually connects the real observable entities, i.e. the substances that can be studied in the laboratory. Such paradigmatic change, in which *ad hoc* hypotheses are not more in *the focus*, is in principle the orientation to *new scientific epistemology*. However, as I will show in the following lectures, *ad hoc* hypotheses still play an important role in science.

## 8.1 Conclusions

Reexamination of phlogiston theory by Lavoisier is in the base of new paradigm in chemical science. It is interestingly how the discovery of only one element, oxygen, has triggered scientific revolution in chemistry by establishing following fundamental concepts of this science: definition of chemical element, chemical classification based on the combinations of elements in the formation of real substances (chemical compounds), and by designing the rational chemical language, the chemical nomenclature. Such system of concepts has opened door for the construction of the chemical structural theory in the near future.

## Reference

1. Lavoisier AL (1965) Introduction: McKie D Traité elémentaire da chimie, Paris (reprint, Dover, New York

# 9 Atoms and Molecules

*Either mathematics is too big for the human mind, or the human mind is more than a machine.*

Kurt Gödel

**Abstract**

This chapter is rich with new ideas about general questions which can emerge from chemical reasoning; from the relationship between thinking in logics and thinking in analogy, till the problem of identity. The atomic theory is discussed as the consequence of quantitative approach in chemistry, i.e. how it is possible to experimentize with invisible microcosm.

**Keywords**

*Ad hoc* hypotheses · Transduction · Law of conservation of mass · Problem of identity · Chemical composition · Equivalence of masses · Atoms · Dalton's laws · Chemical dualism

Ladies and gentlemen,

As I tried to represent in the previous lecture, scientific revolution was also an epistemological revolution which (partially!) has removed *ad hoc* hypotheses from the fundamental basis of theory. In contrast to the phlogiston theory, some of *ad hoc* hypotheses were transformed in the theories about the real physical objects. The *ad hoc* hypothesis in the time of Democritus and Leucippus has been the concept about the atoms as bearers of the discrete structure of matter. Both the philosophers were the opponents to the Aristotle's idea about the continuous substance. In spite



H. Vančik, *Philosophy of Chemistry*, Integrated Science 2,
https://doi.org/10.1007/978-3-030-69224-7_9

to the prevailing Aristotelian picture about the world during the long period of the western science, the concept about atoms did not disappear in the whole history of science: it has survived from the time of Lucretius and Pierre Gassendi, untill the works of Robert Boyle and Ruđer Bošković. However, atoms were still a sort of *ad hoc* category, as well as the Stagiritia's representation of the continuous matter was also an *ad hoc* concept. For the reconstruction of the theory about atoms from the *ad hoc* notion to the physical reality, the support from the experimental observations was necessary. The relationship between an abstract idea and the real experimental observation is in epistemology known as *transduction*. **John Dalton** (1766–1844), known as the father of atomism, and who already was the representative of the new science, has exposed transductive method in his quantitative experiments. Quantitative approach has redirected Dalton to new ideas about atoms as real particles that have their masses as the measurable property.

*Ceteris paribus* experiments performed in closed vessels, i.e. in the isolated systems, led Lavoisier to the discovery of the *law of conservation of mass* in 1789. The formulation of the law of conservation of mass is, however, only the beginning of the quantitative chemistry, which was after that oriented to the measurable relations, primarily the relations of masses and volumes of the reacting substances during the formation or decomposition of chemical compounds.

Remember, that already Stagiritus talks about the two ways in which the substances can combine: compound, μιξις (*mixis*) , and mixture, συνθεσις (*synthesis*). It is important to point out, that the notions such as *mixture* and *chemical compound* were not completely diversificated even in the work of Lavoisier. In that time, it was actual the opinion that the gases nitrogen and oxygen are in the air loosely bound to each other. Thus, while chemical elements have already obtained their *identity*, the identities of chemical compounds were still undefined. While chemical element is defined as a substance that cannot be chemically decomposed to simpler matter, chemical compound should be regarded as a substance that can be *transformed by chemical reaction*, and not just isolated from the mixture.

If this is correct, then chemical compound must possess the equal universal *identity* characterized by exactly defined properties, *independently on the origin* of this compound, or, whether this compound is in the mixture with other stuff or not. The most basic property that can be measured is the ratio of masses of elements from which the compound is formed, i.e. its *composition*.

From the first lecture, we know that Kant has stimulated his student Jeremias Richter to investigate systematically the rations of masses of the substances that react to each other. Convenient experimental model for this study was the neutralization of acids with bases. Richter has found that the masses of acids and bases that form salts by neutralization are in the constant ratios to each other [1]. **Ernst Gotfried Fischer** (1754–1831) has reorganized Richter's data by using the *mass* of sulphuric acid *as a reference*. Since these are reactions of acids with bases, Fischer has designed the table in which the weight proportions of various bases that are necessary for the neutralization of the 1000 weight units of sulphuric acid are represented. These masses are *equivalent* to the mass of 1000 mass units of sulphuric acid. The conclusion is evident, the substances, chemical compounds, react mutually in the exact ratios of their masses.

The mass ratios of the substances which by the mutual reaction yield new compounds has systematically studied **Joseph Louis Proust** (1754–1826), but using the simplest, more rudimentary model. Proust has calculated the ratios of masses of oxygen and metal (iron) which by reaction from various oxides. The result was surprizing: similarly as it was found for neutralizations of acids with bases, **chemical elements, which by reaction form compounds, combine in the definite ratios of their masses**, the formulation that is known as *Proust law*. Although well supported by experimental arguments, the law has not been momentarily accepted in public. **Claude Louis Berthollet** (1748–1822), who is known for the discovery of the chemical composition of ammonia and for the early concept of chemical equilibrium, has rejected the idea about the constant mass rations. At that time it was difficult to oppose the authority of Berthollet. Proust law was finally accepted in 1811, when another authority, Jöns Jacob Berzelius, the scientist about whom I will talk in the following lectures, has confirmed its validity.

If the exact elemental *composition* would be the criterion for the definition of the *identity of chemical compound*, then particular chemical compound must have the same composition independently where it is found, or by which procedure it is prepared. Thus, its identity is independent on the way of its formation. Say, it is the *first order identity (primary identity)*. Chemical elements and compounds possess their first order identity, which is *independent on the history* of their formation. If a substance, or any other object obtains and changes its form during the irreversible history, it has its *individuality*, the *second order identity (secondary identity)*. While, theoretically, the number of entities with the same first order identity could be infinite, the second order identity belongs to only one entity, or to only one object. For instance, *crystal is such an object that possesses the second order identity that emerges from its unique morphology, which is a consequence of the history of its crystallization and growth*—there are no two identical macroscopic crystals. Thus, chemistry can clarify also the metaphysical categories such as identity!

Proust has shown, that the compounds with the same elementary composition (i.e. which have their first order identity) can appear by quite different pathways: for instance, basic copper carbonate, which Proust has prepared in laboratory, has the same composition and properties as the same substance found in nature, the mineral malachite. How to explain the findings of the constant elemental composition? Is it a consequence of the structure of microcosm?

The explanation of the nature of air has required new quantitative investigations. Fortunately, the study of gasses includes more measurable parameters: mass, volume, pressure, and temperature. First laws that describe gasses has already described Robert Boyle. Special interest for gasses has expressed John Dalton during his investigations of meteorological appearances in the atmosphere. Dalton has found, that the composition of air does not depend on the altitude above the sea level in spite to the fact that oxygen and nitrogen have different densities. It would be expected that the denser gas prevails in lower atmospheric layers, and the less dense gas in higher layers.

To explain the constant composition of air, Dalton has used the old-standing idea about atoms as a heuristic concept [2]. Dalton has immagined atom as an indivisible particle of chemical element surrounded by the "mantle of heat". Undoubtedly, the heat that Lavoisier has included in elements, is for Dalton the component of the atom, its mantle. The heat mantle is for Dalton the cause that the forces between equivalent atoms are repulsive, and between different atoms the forces are attractive. Thus, it is expectable that oxygen and nitrogen would tend to mix rather than to separate in layers, since their atoms are different, and they attract each other. The concept about the attractive forces between different, and the repulsive forces between equivalent atoms is one of the origins of *chemical dualism*, the idea that has been prevailing through the longer period of the history of chemistry. These studies led *Dalto*n also to the formulation of the *law about the sum of partial pressures*: the total pressure of the mixture of gasses is the sum of the partial pressures of all its components.

His atomic theory Dalton put in the relation with the real measuring of macroscopic appearances [3]. Dalton has extended Proust concept by studying chemical substances which have the same elements, but combined in different ratios of their masses. For instance, nitrogen and oxygen form a series of oxides in which the oxygen/nitrogen mass ratio is different. From these studies Dalton has articulated the known *law about the multiple ratios of masses*. This law has additionally supported the theory about atoms, the theory that helped to explain the observation that in various nitrogen oxides exists the various number of oxygen and nitrogen atoms. For the first time in the history, the world of atoms, the microcosmos, has been *represented by models*.

By reaching out to the methods of Richter and Fischer, who have already introduced the equivalent mass relationships, Dalton has derived quantitative systematization of chemical elements. Dalton defines the equivalent mass of elements by using the mass of hydrogen, the element with the smallest weight, as a reference. If the reference mass of hydrogen is equal to 1, then the relative "atomic mass" of the element bound to hydrogen can easily be calculated from the known weight ratio of particular element and hydrogen. Within the accuracy of measuring in that time, Dalton has calculated equivalent masses (today we call it "relative atomic mass") of known elements. For instance, 7 for oxygen, 6 for nitrogen, and 5 for carbon.

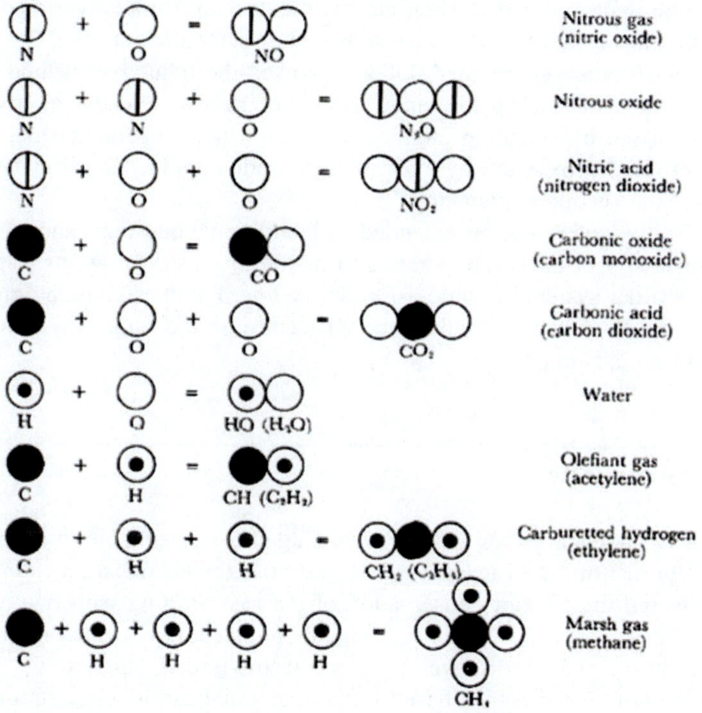


**Dalton symbols of atoms and their combinations (the figure is copied from public internet pages)**

In effect, it means that 7 grams of hydrogen combines with 1 gram of oxygen, or 6 grams of hydrogen combines with 1 gram of nitrogen. Thus, atom of oxygen is seven times heavier than the atom of hydrogen, etc. Since these values are far from the relative atomic masses that are known today, their use in the analyses of a larger number of substances often led to contradictory conclusions. This would motivate Karl Popper to reject the theory about atoms. However, it was this contradiction that became the basis for the development of structural theory, the topic of the following lecture.

There emerges also another question. If Dalton would publish his work by using todays criteria instead the criteria of the beginning of 19th century, he would never discover atoms! If the weights of elements that combine to form the compounds would be measured with the high accuracy, then no integer numbers of weights ratios would be observed. This is important fact about the nature of the scientific method. The discoveries, as well as scientific laws and concepts, are the consequence of the historical state of the art in the moment or their discovery. Historical circumstances are inseparable part of epistemology.

Thus, the appearance of Dalton atomic theory was stimulative for our considerations about the nature of the formation of new scientific concepts and ideas. It is a matter of speculation whether Dalton has used the inductive method, i.e., from observation and measuring to atomic theory, or he was already, in advance, fascinated with the old-standing atomic theory as a heuristic model that should be either confirmed or rejected by experiments. Undoubtedly, Karl Popper would be inclined to the second explanation.

After Dalton, the science is extended with additional questions and uncertainties: Are atoms a sort of *ad hoc* hypothesis? Has the discovery of atoms re-established the Democritian system? Is phlogiston an *ad hoc* hypothesis? Is the discovery of electron, about which I will talk in the XI lecture, in the same way vindicated the idea about phlogiston, or not?

## 9.1 Conclusions

Quantitative chemical experiments have changed the epistemology of this science. By accurate measuring of masses before and after the chemical reaction, Lavoisier has discovered the Law of conservation of masses. Starting with two quantitative laws that he discovered, Dalton has been able to take a look deeper in the nature of the microcosm, and to discover its atomic structure. Both, atomic theory and quantitative laws represent the base for the conceptualization of the future structural theory. Analyzing the concepts of atoms that can combine in complex substances with corresponding elemental composition led us to new approach to the more general philosophical question about the identity problem.

## References

1. Szabadváry F (1962) The birth of stoichiometry. J Chem Ed 39:267
2. Rocke AJ (1984) Chemical Atomism from Dalton to Cannizzaro, Columbus.
3. Needham P (2000) Atomic notation and atomistic hypotheses. Found Chem 2:127–180

# Systematization, Classification, Structure, and Elements

# 10

*There is nothing either good or bad but thinking makes it so.*
William Shakespeare

**Abstract**

The new definition of the notions of structure and element is in the focus of this chapter. It is explained how the chemical epistemology based on analogies and classification can led to the "invention" of the structure, and to its representation by chemical formulas and chemical nomenclature. Structural theory that origins from the theory of radicals and the theory of types is in the base of modern chemistry. I intend to show how the periodic law of classification of chemical elements is constructed by using the same methodology that was successful in the formation of the structural theory.

**Keywords**

Analogy as method · Structural theory · Structure · Isomorphy · Composition · Constitution · Isomerism · Vitalism · Epistemology of contents-language-script · Theory of radicals · Theory of types · Critics of radical falsificationism · Periodic system of elements · Concept of valence · Chemical synthesis · Concepts of quality and quantity

Ladies and Gentlemen,

Dalton's law about the multiple ways of combinations of atoms in chemical compound represents, in principle, the difficulty in experimental determination of equivalent atomic masses. Since only the **ratio of weights** of two elements, which



H. Vančik, *Philosophy of Chemistry*, Integrated Science 2,
https://doi.org/10.1007/978-3-030-69224-7_10

form chemical compound, can be measured, and since the **ratio of atoms** of elements on microscale is unknown, the calculation of atomic mass can be contradictory. For instance, Dalton has found that 87.4 weight ratios (in grams or in other units) of oxygen combines with 12.6 weight ratios of hydrogen. From these data, it follows that the ratio of hydrogen and oxygen is 1:6.94. If it is taken that the relative atomic mass of hydrogen is 1, than the relative atomic mass of oxygen should be 7 (from the measured value 6.94). However, this relative atomic mass of oxygen (7) is true only with the presumption that water is formed from only **one** atom of oxygen, and only **one** atom of hydrogen. In modern words, the formula of water would be HO.

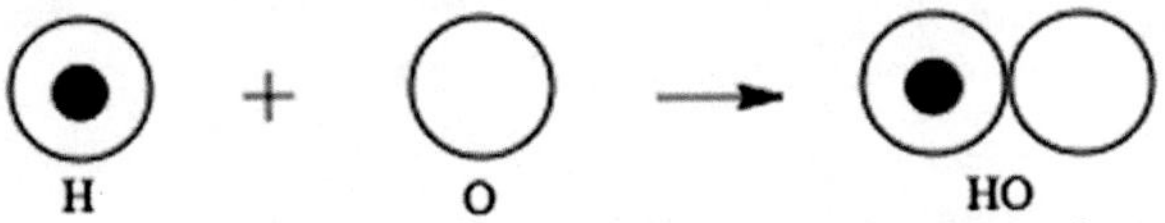


Conversely, if water is formed from another combination of atoms, say $H_2O$, oxygen should have relative atomic mass 13.9!

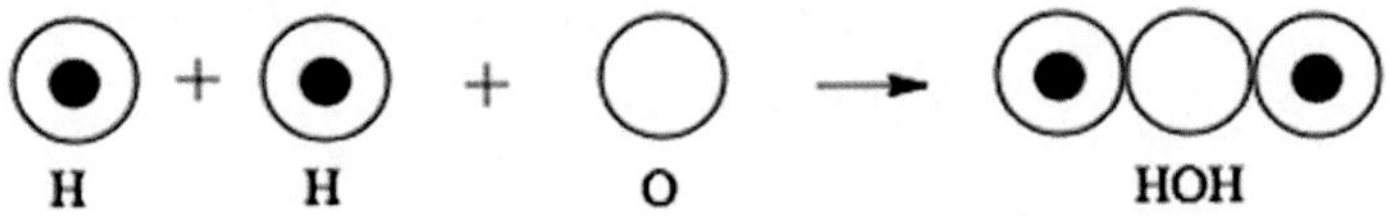


In conclusion, for the determination of the relative atomic masses of elements that combine in compound we need two parameters: their mass ratio, and the number of atoms of particular element in combination, respectively. However, the gravimetric experiments afford only one parameter, their mass ratio! There is simple mathematics: it is not possible to calculate two variables from only one equation!

Represented epistemology is evidently insufficient, and it requires new paradigmatic solution. This new extension is metaphysically a sort of return to old (al)chemical tradition: **thinking in analogies**. For example, if water and hydrogen sulphide are analogous, and if we know from the experience that hydrogen sulphide has the ratio of atoms $H_2S$, than water should have the same ratio of atoms, $H_2O$, and the relative atomic mass of oxygen is (from Daltons measurements) 13,9. In contrary, if the ratio in hydrogen sulphide would be HS, and water is analogous to it, than the atomic mass of oxygen would be 7.

Thus, determination of the equivalent atomic masses of elements is possible by using the new, perhaps vague, category—*the analogy*. What is chemical analogy at all? How to explain it clearly, and are there any unequivocal empirical parameters? These questions have redirected the development of chemistry and science to one new fundamental concept—*the structural theory* as a basis for the representation of microcosm.

The idea about the chemical structure and the idea about analogies were mutually intertwined. Why are hydrogen sulphide and water analogous in spite to the fact that these substances have different properties? Is there a rational criterion for analogy? Important step in the solution of this question has happened within the research group of Swedish chemist **Jöns Jacob Berzelius** (1779–1848). His collaborators **Friedrich Wöhler** (1800–1882) and **Eilhard Mitscherlich** (1794 1863) have represented both the concepts, analogy, and structure, respectively, in the new light. During his investigations of the morphology of crystals, particularly the crystals of phosphates and arsenates, Mitscherlich has discovered that the substances different in elemental composition can form the crystals with the same form and symmetry. Mitscherlich has called it *isomorphy*. Thus, the *structure*, which is, in this case, observed as a morphology of crystal, is independent on the composition of the substance. *Structure is an autonomous phenomenon* that could be used as possible criterion for analogy. Thus, the substances are analogous if they form the crystals with the same morphology and symmetry. The simplest examples of isomorphy are represented in following figures. Since sodium-chloride forms the crystals with the same form as crystals of potassium-bromide, sodium-chloride and potassium-bromide are analogous. If they are analogous, than the ratio of sodium atom and chlorine atom should be the same as the ratio of potassium atom and bromine atom, and it is 1:1. Then, if we know the relative atomic masses of sodium, potassium, and chlorine, and the experimentally determined mass ratio of potassium and bromine, the relative atomic mass of bromine can simply be calculated.

**Crystals of potassium bromide KBr (the figure is copied from the public internet pages)**

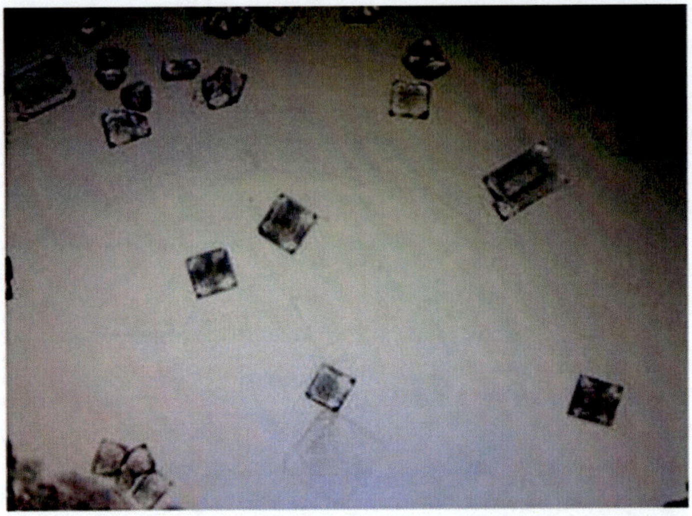

**Crystals of sodium-chloride NaCl (the figure is copied from the public internet pages)**

Friedrich Wöhler is in history of chemistry famous with its experiment (1852) in which he succeeded to prepare organic compound, urea, from the inorganic precursor ammonium-cyanate:

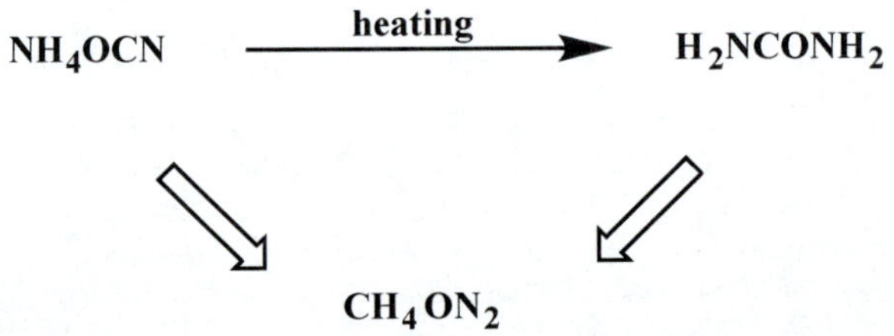


Until that experiment, the chemical theories were mostly focused (and valid) to inorganic compounds. Behaviour of compounds isolated from plants or animals has been unexplainable within the old theories by Stahl and Lavoisier. In his book about chemistry, *Systéme des connaissances chimiques* (1800) **Antoine Fourcroy** (1755–1809) describes organic compounds only in the last two chapters. However, just this orientation to the systematic investigation of this class of compounds will represent the key step for the final development of structural theory.

Far-reaching of Wohler's discovery is not only in the rejection of dualistic hypothesis by which the chemistry in living beings is different from the chemistry in inorganic world, but also in the knowledge that the same elemental composition (represented by formula $CH_4ON$) can possess very different substances. The conclusion is that, if the difference between substances is not only in their composition, it must be in something else: the micro organization of atoms that we call *the structure*. Works of Mitscherlich and Wöhler are complementary, while Mitscherlich has demonstrated that different substances can have the same structure (observed as the same form of crystals), Wöhler has shown that the same elemental composition can have substances with their different structure. Such an abstract notion of structure has been reached by starting from two independent approaches.

However, the category of structure has slowly transformed in more realistic form, i.e. it was re-defined as a way in which atoms are interconnected with each other: *the constitution*.

It could be speculated, that Berzelius has possibly in mind these experiments when he introduced the notion of *isomer*, the substance that possesses the same elemental composition, but the different constitution.

Thanks to Berzelius, two new important concepts of structural theory, *composition* and *constitution*, were introduced in chemistry. Composition, the quantitative elemental content of the measured mass ratios of elements, determines the number of atoms of particular element in the compound in agreement with the Dalton's theory of atoms. However, as I have shown in the example with water, the accurate composition is possible to determine only if we have the exact values of the relative atomic masses. Application of Mitscherlich analogies was limited with the relatively small number of isomorphic crystals. The solution of this problem is either in finding new criteria for analogy, or to obtain the relative atomic masses in another way. Berzelius has succeeded to determine the relative atomic masses relatively accurately, within the possibilities of measuring at that time. As a reference value, Berzelius has used the mass of oxygen instead the mass of hydrogen. Indeed, the Berzelius values are not very far from the relative atomic masses that we use today.

Already from phlogiston theory and Lavoisier revolution, the problem of the substances of animal and plant origin has remained unsolved, because none of these theories were able to comprise this question. Particular chemical behaviour of the substances of plant and animal origin has been reflected in the emergence of the *vitalistic thinking*, by which chemistry in living systems is different from the chemistry in inorganic world. Jöns Jacob Berzelius, as a convinced vitalist, has deserved his work to chemistry of organic compounds. His experience in experimentation with the compounds isolated from living beings Berzelius has applied to the systematic examination of elemental compositions of organic substances. Since organic substances burn by producing water and carbon dioxide, Berzelius has succeeded to calculate compositions of organic compounds from the measured masses of these products of combustion. Starting with the atomic masses, which he has determined, Berzelius has not only calculated compositions, but also developed the symbolism that represents it. As symbols for elements, Berzelius has used one or two letters of their Latin names, and the number of atoms in the chemical compound is labelled with the numeric indexes above the letter:

**H²O**, for water (two atoms of hydrogen and one atom for oxygen), or **CH⁴** for methane (one carbon atom and four hydrogen atoms) the symbols for elements are: **H**ydrogen, **O**xygen and **C**arboneum.

Chemical elemental contents of a compound, its composition, is in this way represented by exact symbolism—the chemical formula. **Justus Liebig** (1803–1873) has changed this symbolism in such a way that the numerical label of number of atoms became bottom index: $H_2O$ or $CH_4$. Some chemical books and textbooks published in the first decade of 20th century still have Berzelius symbols. Soon I will explain that chemical formulas are not only iconic labels for particular compounds, rather, they have much broader informative contents about the substances. Similarly, chemical nomenclature developed in the Lavoisier's circle was not only a collection of names and symbols.

Besides the **contents**, i.e. substances and transformations, the chemical conceptual system is complemented with special **language** (nomenclature), and special **script** (formulas). This *epistemology of contents, language, and script* has its origins already in the beginning of chemical history—in protochemistry, as it has been discussed in the 3rd lecture represented in this book.

By systematic determination of composition of organic compounds, Berzelius has observed that some weight ratios of carbon and hydrogen (some ratios of the numbers of carbon and hydrogen atoms) appear repeatedly, for instance compositions $CH_3$, or $C_2H_5$, or $C_6H_5$. It seemed as organic compounds are merely the combinations of these groups of atoms. As an example, the composition $C_2H_6$ seems to be composed from two groups of $CH_3$, or $C_7H_8$ is a combination of one group $C_6H_5$, and one $CH_3$. Berzelius was convinced that these groups are stable, and that in organic compounds they play the same role as chemical elements play in the inorganic compounds. For these groups, Berzelius has used the term *radicals*. Organic substances are, thus, decomposable in radicals, i.e. in polyatomic fragments, which than could be combined by similar principles as atoms of elements combine to form compounds. In such a way, this *theory of radicals* begun to create the first concept of structural theory:

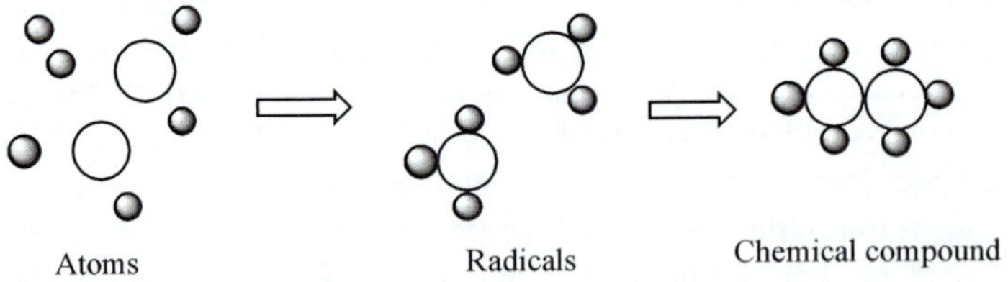


However, to be developed in the conclusive theory, the idea about the structure, which has appeared in the works of Wöhler, Mitscherlich, and Berzelius, requires another additional step—the extension of the category of analogy.

Morphology of crystals as a basis for the determination of analogy was too narrow. For the establishment of the unequivocal reference for analogy, the appearance of new paradigm was essential. If the criterion for analogy cannot be determined by the immediate observation of properties of the substance, perhaps the better approach can be found in focusing on the changes of substances, i.e. on chemical reactions.

But, before disputing the ways through which the concept about the structure can emerge from the systematic investigation and classifications of chemical reactions, we have to look back on another important discovery, which has provided a new light on the question of chemical affinity, and the problem of mutual binding of atoms.

By the invention of the first battery, and discovery of the electrical current, **Alessandro Volta** (1745–1827) has initiated new field of chemistry—the *electrochemistry*. Volta has stimulated the development of methods by which chemical changes can be triggered by application of electrical current. By electrolysis of water, Volta has, again after Lavoisier, demonstrated that water consists from oxygen and hydrogen, the gases that are produced on the different electrical poles, hydrogen on negative, and oxygen on the positive pole. This experimental approach has appeared as new chemical method, chemists started to electrolyse all available substances. **Humphry Davy** (1778–1829) has electrolyzed the melts of salts at high temperature, and he succeeded to isolate a series of new chemical elements, metals *sodium*, *potassium*, *calcium*, *strontium*, and *magnesium*. All these metals have emerged on the negative electric pole (i.e. the negative electrode), the same as it was the case with hydrogen. Oppositely, non-metals have always developed on the positively charged electrode. *Chemical affinity*, or we could call it already the *chemical bond*, is by its nature an electrical phenomenon! Clearly, atoms of metals can combine with atoms of non-metals because their electrical nature is, as it follows from the electrolysis, opposite. This is one of the origins of the first theory about chemical bond, the *chemical dualism*: only atoms of different elements can be bound to each other! Berzelius, who had performed a series of electrochemical experiments, has insisted on this dualistic view.

However, preparation of series of chemical compounds, especially organic, has provided the facts that were not explainable within the dualistic point of view. **Jean Baptiste Andre Dumas** (1800–1884) has prepared new organic compound, trichloroacetic acid, by irradiation the mixture of chlorine and acetic acid with the sunlight. In modern formulas the reaction can be represented as follows:

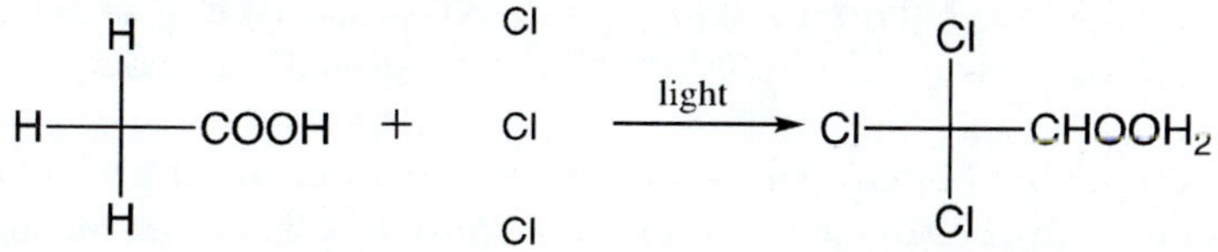


In this transformation, the structural change that has been not explainable within the dualistic theory has been observed: hydrogen, which develops on the negative electrode, is in this reaction replaced with chlorine, for which it is known that appears on the positive electrode. Thus, elements, their atoms, with the opposite electrical nature are mutually substituted. Consequences of this discovery are far-reaching. First, since this is the discovery of a new type of chemical reaction, the *substitution*, the chemical theory has introduced new way of *classification, classification by the type of reaction.* Chemical classification became "two-dimensional", in one direction is the classification of substances, and in another direction is classification by type of reaction, similarly as I have been anticipated in discussions about the Newton's elective affinity (in the fifth lecture of this book). The second consequence is the possibility that the substitution reaction can be extended in more general concept. In the simplest case, in hydrogen chloride, any of these two elements, hydrogen or chlorine, respectively, can in principle be substituted with any atom of different elements. For instance, hydrogen is replaced with sodium, and chlorine with bromine:

$$\mathrm{HCl} + \mathrm{Na} \longrightarrow \mathrm{NaCl} + \mathrm{H}$$

or

$$\mathrm{HCl} + \mathrm{Br} \longrightarrow \mathrm{HBr} + \mathrm{Cl}$$

More generally, we have obtained the conceptual model in which two atoms are substitutable with any of other atoms:

This universal two-atomic combination Dumas has called the *structural type*, in this case the *structural type of HCl.* Structural type, the elemental entity of the structure is *new category*, which is independent on the actual chemical composition. The idea about the structure as a concept that defines the way how atoms are mutually interconnected, which has been already anticipated in works of Wöhler and Mitscherlich, has appeared in the focus of discussions in chemistry.

Although Dumas must be considered as an initiator of the theory of types, (which some historians call the old theory of types) its final form was designed by **Charles Frédéric Gerhard** (1816–1856). In this new theory, which Gerhard has published 1853, the four structural types were proposed: *type of hydrogen chloride,* HCl, type of water, $H_2O$, *type of ammonia,* $NH_3$, and *type of ammonium*, $NH_4$ (today it is $NH_4^+$) which August Kekulé has later replaced with the *type of methane*, $CH_4$.

By substitution of atoms that constitute water with atoms or group of atoms of other elements (hydrogen atoms, or oxygen atom) we obtain the structural type of water that have been introduced by **Alexander William Williamson** (1824–1904). Similarly, type of ammonia (**Charles Adolph Würtz**, 1817–1884), as well as the type of methane were created in similar way. It must be pointed out, that although these structural types are not the representations of molecular structure in modern sense, they were an important step in the development of chemical theory. This could clearly be seen from the symbolism by which the types are drawn from the knowledge about substitution reactions:

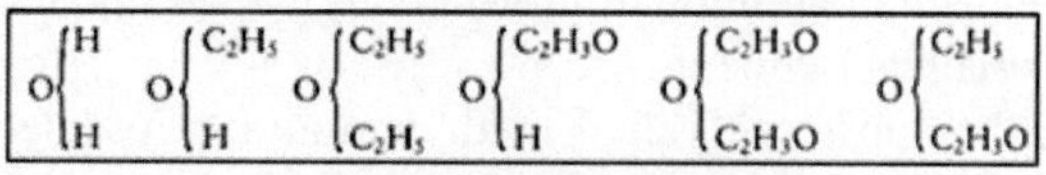

TYPE OF WATER

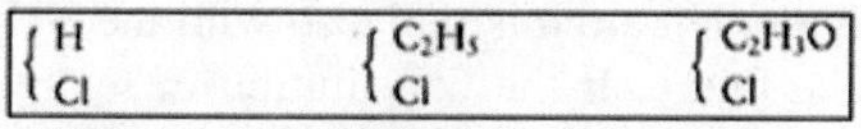

TYPE OF HYDROGEN CHLORIDE

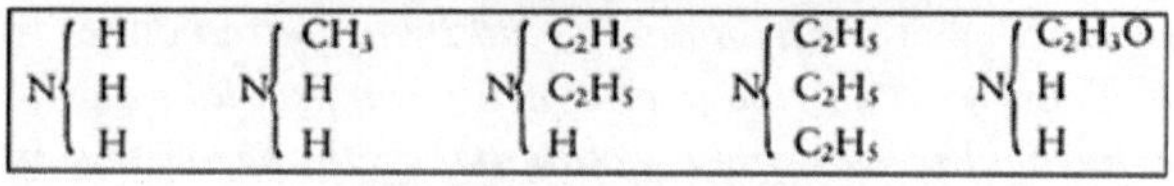

TYPE OF AMMONIA

On the right side of the curly bracket are atoms or groups replaceable in substitution reaction. The structural type is in this substitution preserved. In modern symbolism these types could be represented in following way:

**Structural types**

Type HCl HCl, NaCl, KBr,…

Type $H_2O$ $H_2O$, $H_2S$, $(CH_3)_2Se$, $Na_2O$, ..

Type $NH_3$ $NH_{3,}$ $(CH_3)_2NH_{2,}$ $PH_3$,..

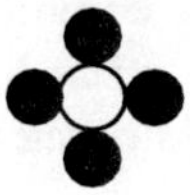

Type $CH_4$ $CH_{4,}$ $CH_2Cl_{2,}$ $SiCl_4$,..

This first structural theoretical concept has as a consequence the extension of our information about chemical compounds. The substance is now represented not only with its *composition*, i.e. the number of atoms of particular element, but also with its *constitution*, i.e. the manner how atoms are mutually interconnected. From constitution we have also an additional information, namely, how the structure can be decomposed in smaller fragments.

For instance, the substance called methylamine (represented with modern formula as $CH_3NH_2$) belongs in the same time to two types, to the type of ammonia in which one hydrogen atom is replaced with the $CH_3$ group, and to the type of methane where one hydrogen atom is replaced with the $NH_2$ group. Both the types are recognizable in the figure. If the constitution of methylamine could be represented as a combination of two structural types, ammonia and methane, respectively, than it follows that *methylamine could be prepared* either from ammonia, or from methane! Thus, *our knowledge of constitution provides new information about the way how particular substance could be synthesized*: this is a paradigmatic change because extends chemistry *from analytical to synthetical discipline*! Chemistry became not only science, but also a sort of architecture of new unknown substances. Radicals, proposed by Berzelius are also reducible on structural types.

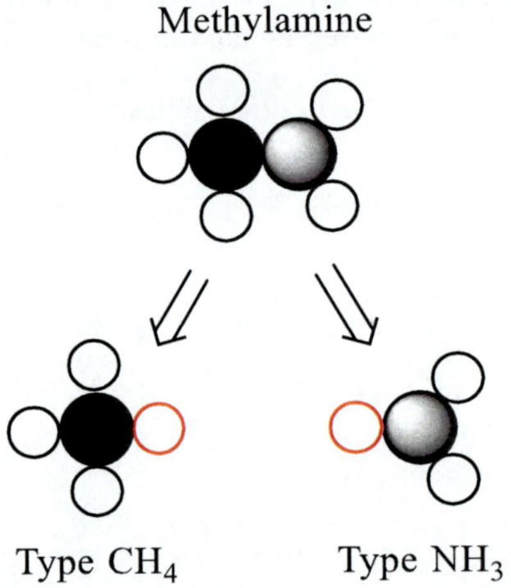


Berzelius "structural theory of radicals" has been, in principle, contradictory to his own dualistic statements that only the different atoms can combine to form chemical compounds. How it is possible that two radicals, for instance with elemental composition $CH_3$, can led to ethane, $CH_3CH_3$, if two carbon atoms (which have the same nature) are not bound to each other?

To resolve this contradiction, we have first to talk about the new concept that emerges from the type theory—*the valence*. By systematic investigation of the chemical syntheses of alcohols and ethers, the compounds classified to the structure type of water, Williamson has introduced the notion of valence. For him the structural type of water is *bivalent*. Analogously, the types of hydrogen-chloride, ammonia, and ammonium are monovalent, trivalent, and tetravalent. Valence is in such a way defined as a ratio of units, atoms or Berzelius radicals, which are exchanged between each other in substitution reactions. But, we have to be careful: this concept is still not the explanation of chemical bond and its nature, it is merely a measure of quantitative relationship between substances.

**Archibald Scott Couper** (1831–1892) has extended the type theory to structures of organic compounds by the discovery of the *tetravalence of carbon*. After the discovery that carbon atoms can bind to each other and form longer chain structures, Couper has also corrected Berzelius structural theory of radicals by rejecting dualism. The represented examples of Couper symbols for the simplest organic acids (formic acid, acetic acid, and propionic acid) reproduced from his paper *Sur une nouvelle théorie chimique* (1858) clearly show the mutual binding of carbon atoms. The lines which connect the symbols of carbon and hydrogen could already be regarded as a prototype of the constitution structural formula, which are similar to those that are in use today.

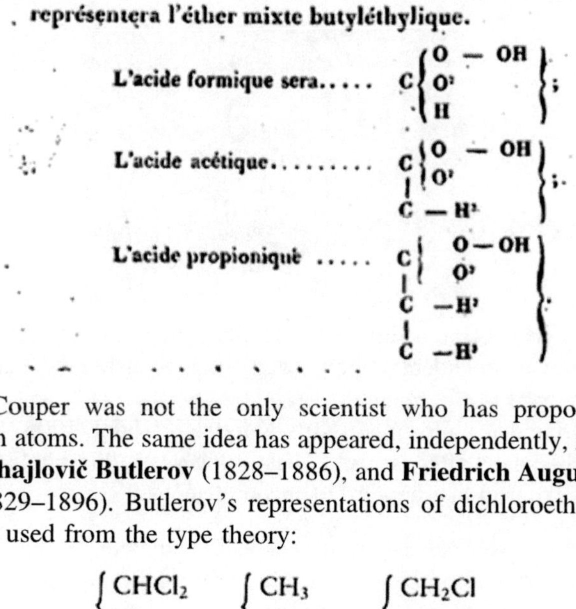


However, Couper was not the only scientist who has proposed the bonds between carbon atoms. The same idea has appeared, independently, in the works of **Aleksandr Mihajlovič Butlerov** (1828–1886), and **Friedrich August Kekulé von Stradonitz** (1829–1896). Butlerov's representations of dichloroethane are still in the symbolism used from the type theory:

$$\begin{cases} CHCl_2 \\ CH_3 \end{cases} \quad \begin{cases} CH_3 \\ CHCl_2 \end{cases} \quad \begin{cases} CH_2Cl \\ CH_2Cl \end{cases}$$

Kekulé has the discovery about carbon-carbon bonds published 1857, one year before Couper [1].

ANNALEN

DER

CHEMIE UND PHARMACIE.

CIV. Bandes zweites Heft.

Ueber die s. g. gepaarten Verbindungen und die Theorie der mehratomigen Radicale; von *Aug. Kekulé.*

**Facsimile of the front page of the word of August Kekulé about mutual binding of carbon atoms in organic compounds**

Scottish chemist **Alexander Crum Brown** (1838–1922) has changed the representation used in type theory by introducing the lines which connect symbols of elements—modern *chemical formula* is born:

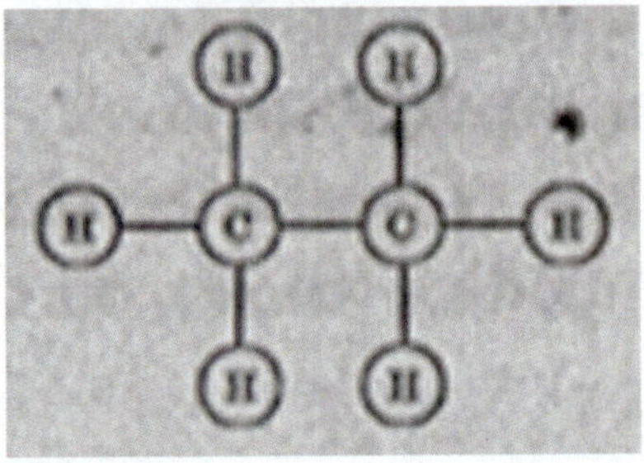

As I have demonstrated in a series of examples, theory, or concept, as well as any logical system, independently how well it is constructed, always finish in contradiction.

Mathematician and philosopher **Kurt Friedrich Gödel** (1906–1978), one of the most famous logicians after Aristotle, has in the year 1931 published the paper in which he has exposed his theorem of incompleteness:

1. If the system is self-consistent, it cannot be complete.
2. Consistency of axioms, which are the basis of the system, cannot be proved within this system.

This revolutionary discovery has changed not only science but also the general way of thinking: it is not possible to create any scientific system that is complete in a way that it never falls in contradiction. Thus, science became more and more a sort of the "organized scepticism", rather than a method for searching the final truth. Although, at first sight, *Gödel theorem* seems pessimistic and discouraging, in fact, the incompleteness of science is the main force that stimulate its development.

Let remember my first lecture where I have exposed two approaches in the validation of the quality of science. By the principle of induction advocated by representatives of the Vienna circle, especially by Wittgenstein, theory must be provable. The opposite approach proposed by Popper requires that a good theory must be refutable, i.e. it must assume the test that can reject it. By using this Popper principle strictly, and keeping in the mind Gödel's theorem, we are getting into natural conclusion that almost all the theories should be rejected because none of them, in accordance with the incompleteness theorem, can pass its own tests. What would remain of science?

The structural theory, used as a conceptual system developed in chemistry, represents good model for the demonstration of this Gödel-Popper discrepancy. Application of the structural theory based on the principle of valence came across unsolvable contradictions. By studying the structure of the benzene molecule with its composition represented as $C_6H_6$, August Kekulé has proposed the structure in which carbon atoms are mutually connected in the six-fold ring. Since at this moment the Crum Brown formulas were not yet in the wide use, Kekulé depicts benzene in different representations shown in the following figures.

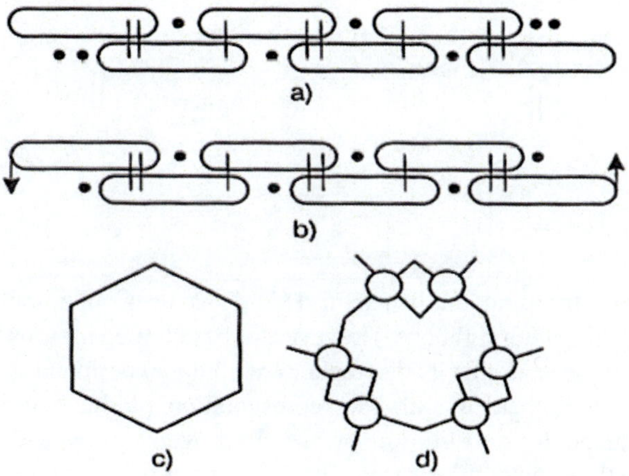


**Representations of the constitution of benzene molecule in different works of August Kekulé**

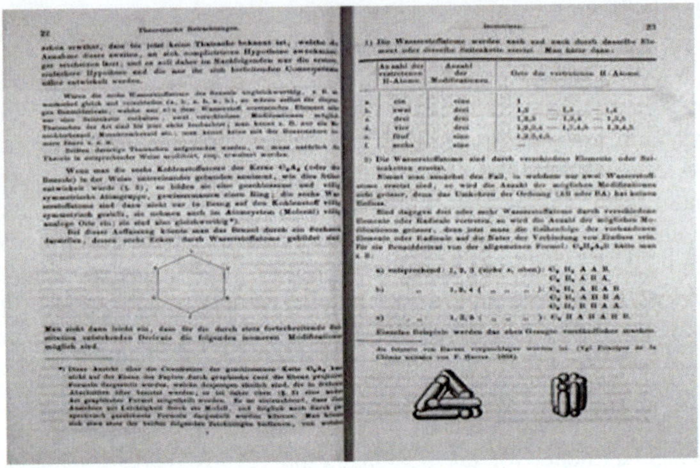

**Facsimile from the dissertation of August Kekulé about the structure of benzene**

In modern notation this structure is described with a six-fold ring with the alternating single and double bonds:

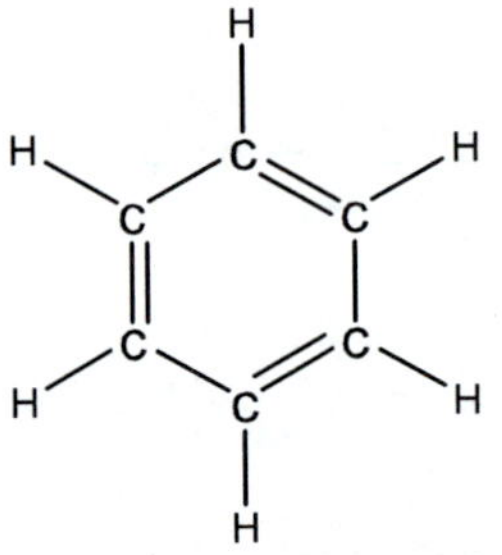

or, simpler

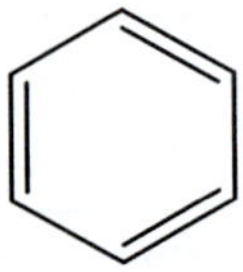

Kekulé ring structure, published in 1856, has been an important step in the extension of the structural theory. However, although elegant, this structural model was questionable because of its discrepancy with the experimental observations [2]. The problem has appeared with the representation of the structures of benzene derivates obtained by substitution of two hydrogen atoms with other atoms or groups, for instance with chlorine. By replacement of two hydrogen atoms with two chlorine atoms we get dichlorobenzene. Starting from the Kekulé structure of benzene, we could expect four different *isomers* of dichlorobenzene, A, B, C i D:

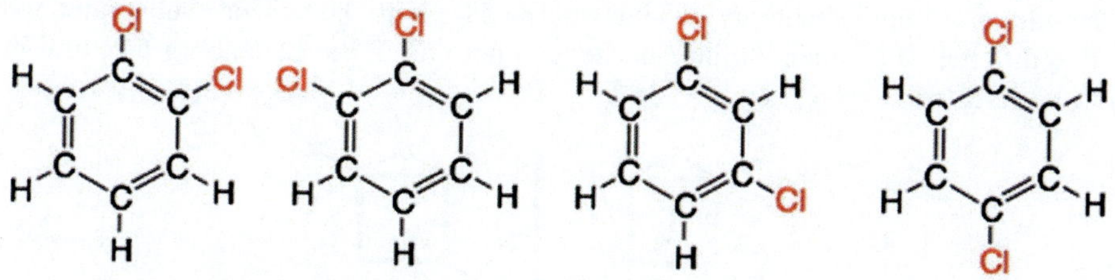

a b c d

Accidently, all the experiments have shown that only three isomers exist, namely, the structures A and B are the identical compound![1] Thus, Kekulé's structural concept was in contradiction almost in the moment when it has appeared. In accordance with Popper criterion, this theory did not passed the experimental test, and it must be declined. However, if this contradictory theory would be rejected, the development of chemistry would be for sure retarded because the benzene molecule is really a six-fold ring of carbon atoms, and this fact was very important in synthetic organic chemistry. In this time, especially in Germany, the newly developed chemical industry was focused on the production of dyestuffs, the material for which benzene is the basic component.

Since Kekulé structure did not passed the experimental test, **Albert Ladenburg** (1842–1911) has proposed the alternative structure that satisfies the $C_6H_6$ formula [3]. In literature, this compound is known as *prismane*. Indeed, if this is the structure of benzene then only three isomers (ortho, para, and meta) appear. In accordance with the Popperian criterion the Kekulé structure must be rejected in favour of prizmane.

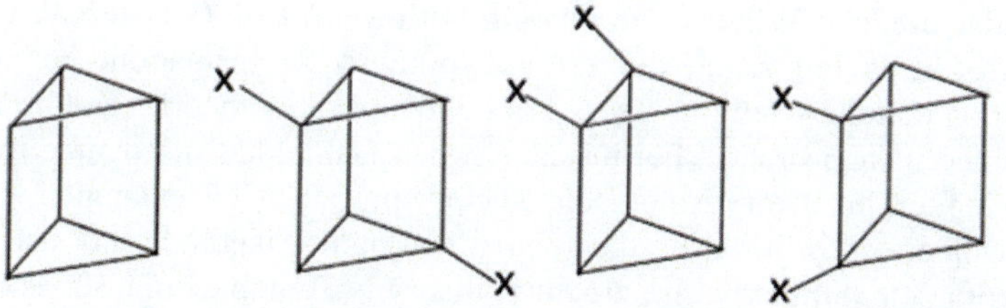

ortho para meta

However, after the discovery of chirality of molecules (*vide infra*), it became clear that prizmane, if substituted with the two different substituents (x and y),

[1]Structures A and B differ in the Scheme between each other in the type of bond between the C-atoms on which is bound Cl. While in the structure A, the bond between these C-atoms is double (C=C), in the structure B, the atoms are bound with the single bond (C–C).

should afford optical activity, what is not the case with the benzene substituted with two different substituents. Since benzene is not chiral, the Ladenburg proposal has to be rejected.

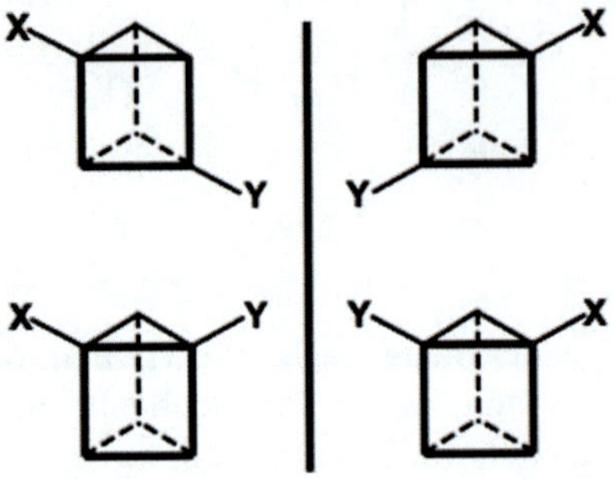


This example demonstrates the weakness of the Popperian criterion. The problem is that Popper does not take into account the *passage of time in the sequence of discoveries*. The discrepancy of Kekulean structure with experiment exists in particular period of time in which Ladenburg solution seems acceptable. But during the longer period, the problematic Kekulean structure has survived, and the Ladenburg's not. Thus, the sole critical experiment cannot distinguish between the acceptable and wrong theories, because theory or concept is a dynamical organism that changes its variation during the time.

How it is possible that the theory of high importance for science is contradictory? Evidently, the structural theory of that time has reached the "Gödel limit" of incompleteness.

Remember my first lecture in which I have discussed the carefulness with the use of the radical version of Popperian criteria. Pierre Duhem, who was mentioned in the first lecture, and **Willard van Orman Quine** (1908–2000) [4] have develop the thesis that it is impossible to test a hypothesis in insulation, i.e. without an assumption that the scientific theory is a bundle of other theories and principles that has enabled its appearance. For instance, when Galilei has discovered the satellites around Jupiter, his theory is not isolated concept, rather it is entangled with a series of other theories, for instance, the geometrical optics, methods and instruments for observation, etc. Similarly, the structural theory is a complex of a series of laws and experiments, as well as theories from other fields, Dalton and Proust law, Lavoisier law of conservation, type theory, Berzelius radical theory, etc. To be relevant, the Popperian test should reject all these theories and laws, not only the structure of benzene as an isolated problem. But then, the science would stop to exist!

Imre Lakatos [5] has extended and relativized Popperian criteria with the requirement that the theory should be *productive*, rather than strictly consistent. By Lakatos, theory is not only a bundle of other theories, laws, and methods, it is an aggregate of *ad hoc hypotheses* which are the driving force for its future development.

Kekulé has also proposed an *ad hoc* theory about the fast scattering of atoms within the molecule. The consequence of these scatterings is that isomers A and B are in very fast interconversion. However, Kekulé did not proposed extra argumentation for this hypothesis.

For the final solution of this contradiction about the structure of benzene we have to wait a half of century. But this solution will appear as a new paradigm, not only in chemistry, but also in the general science.

Now let me discuss some other conceptual solutions around structural theory, which are results of thinking in analogies. Basic idea of structural theory is that molecules can be decomposed in smaller repetitive fragments. Good example are Berzelius theory of radicals. Within this theory Berzelius has proposed new method of classifications of simple hydrocarbons—the *homologous series*. Simple hydrocarbons and their molecules can be classified in homologous series if all have elemental composition that satisfy formula $\mathbf{C}_n\mathbf{H}_{2n+2}$, if $\boldsymbol{n}$ is an integer. Ethane has composition $C_2H_6$, what is in agreement with this formula if $\boldsymbol{n} = 2$. From this regularity, it follows that all these molecules can be composed from identical simple fragments.

Similarly, relative molecular masses can also be decomposed in fragments, numerical fractions. In the following Table are listed the masses and formulas of the simplest organic compounds—the hydrocarbons. Evidently, each of the masses can be decomposed in two fragments of mass 15, and a variable number of fragments with masses 14. Conversely, if masses of any of series of particles can be composed from identical smaller masses, then also the particles as such can be composed from smaller fragments.

| Molecular mass | | | | | | | |
|---|---|---|---|---|---|---|---|
| 30 | $C_2H_6$ | 15 | | | | | $CH_3CH_3$ |
| 44 | $C_3H_8$ | 15 | 14 | 15 | | | $CH_3CH_2CH_3$ |
| 58 | $C_4H_{10}$ | 15 | 14 | 14 | 15 | | $CH_3CH_2CH_2CH_3$ |
| 72 | $C_5H_{12}$ | 15 | 14 | 14 | 14 | 15 | $CH_3CH_2CH_2CH_2CH_3$ |

**Fragments have mass 14, i.e. $CH_2$ group**

Jean Baptiste André Dumas has, inspired with his analysis of the structure of organic compounds in such a way, used the analogous method for the "decompositions" of relative atomic masses of chemical elements. Analogy is again in action! In his work *Mémoire sur les équivalents des corps simples*, published 1858 as one of the editions of French Academy, Dumas has compared differences in masses in particular series of organic compounds with differences in relative atomic masses of chemical elements [6].

( 129 )

**MÉMOIRE SUR LES ÉQUIVALENTS DES CORPS SIMPLES,**

**Par M. J. DUMAS.**

Lu à l'Académie des Sciences le 9 novembre 1857 et le 27 décembre 1858.

Les équivalents des corps simples, c'est-à-dire les poids respectifs des particules matérielles dont la combinaison donne naissance à tous les corps de la nature, ont été déterminés pour la plupart par Berzelius avec une telle atten-

**Facsimile of the frontpage of work of J. B. A. Dumas about mass relations of molecular fragments, and decomposition of relative atomic masses in smaller fragments**

From Dumas original Table it is clear, that the common difference between masses of *Ammonium* and *Méthylium*, from the other members from this group (on the bottom of the Table) is always equal 3.

| | | | | |
|---|---|---|---|---|
| Fluor | 19 | Azote | 14 | Diff. commune 5 |
| Chlore | 35,5 | Phosphore | 31 | |
| Brome | 80 | Arsenic | 75 | |
| Iode | 127 | Antimoine | 122 | |
| Magnésium | 12 | Oxygène | 8 | Diff. commune 4 |
| Calcium | 20 | Soufre | 16 | |
| Strontium | 43,75 | Sélénium | 39,75 | |
| Barium | 68,5 | Tellure | 64,5 | |
| Plomb | 103,5 | Osmium | 99,5 | |
| Ammonium | 18 | Méthylium | 15 | Diff. commune 3 |
| Méthylammonium | 32 | Éthylium | 29 | |
| Éthylammonium | 46 | Propylium | 43 | |
| Propylammonium | 60.. etc. | Buthylium | 57.. etc. | |

**Facsimile of the Table from the work of J. B. A. Dumas about the mass ratios of molecular fragments, and the analogous decomposition of relative atomic masses in smaller components**

This play with numbers led to the same conclusion as in the case of homologous series. For instance, *Propylium* has mass 43, what corresponds to the sum of 15 + 14 + 14, and Buthylium has mass 57, what is sum of 15 + 14 + 14 + 14! I leave it up to you to play with other numbers from the Table.

Thinking in analogies, which is especially characteristic in chemical epistemology, led Dumas to new revolutionary idea. Since the same regularity of mass decomposition exists not only in molecules but also in relative atomic masses, as it is shown in the upper part of the Table, it is expectable that **atoms are also composed from smaller subatomic particles**. The possibility of construction of atom from different sorts of smaller fragments, i.e. the *first epistemic access to the subatomic world, is the discovery that has appeared in chemistry!* As it can be recognized from the Table, the relative atomic masses of elements can be composed from subatomic fragments of mass 16.

| Atomic mass | | |
|---|---|---|
| 7 | Li | 7 |
| 23 | Na | 7 + 16 |
| 39 | K | 7 + 16 + 16 |
| 16 | O | 16 |
| 32 | S | 16 + 16 |
| 80 | Se | 16 + 16 + 16 + 16 + 16 |
| 128 | Te | 16 + 16 + 16 + 16 + 16 + 16 + 16 + 16 |

**Fragments of mass 16, "subatomic group"**

**Composition of relative atomic masses from smaller fragments**

Likewise, **Johan Döbereiner** (1780–1849) has recognized the regularity in the relative atomic masses within the groups of three elements, so called *triades*. Döbereiner has classified elements in groups in such a way that the mass of the element in the middle is the arithmetic mean of atomic masses of two neighbouring elements. Grouping of elements in triades has already anticipated the systematization of elements in groups in the periodic system of elements.

| | | |
|---|---|---|
| Li 7 | Ca 40 | P 31 |
| Na 23 → $\frac{7+39}{2}$ | Sr 87 → $\frac{40+137}{2}$ | As 75 → $\frac{31+122}{2}$ |
| K 39 | Ba 137 | Sb 122 |

| | |
|---|---|
| S 32 | Cl 35.5 |
| Se 78 → $\frac{32+128}{2}$ | Br 80 → $\frac{35.5+127}{2}$ |
| Te 128 | I 127 |

**Döbereiner *triades*. The element in the middle in every group of three elements has the relative atomic mass that is arithmetic mean of masses of upper and lower element**

In the meantime, other authors have systematized chemical elements in different ways starting from atomic mass ratios. **John Alexander Newlands** (1837–1898) has classified elements in eight groups. He has called it a *law of octaves*. **Julius Lothar Meyer** (1830–1895) has, by using the already developed notion about valence, systematized chemical elements by their valence. Since valence is the main property that determines the nature of chemical elements, Lothar Meyer contemplates about the new chemical law about the periodic repetition of behaviour of chemical elements, the law that was in its final form discovered by **Dmitri Ivanovič Mendeleev** (1834–1907).

| I | II | III | IV | V | VI | VII | VIII | IX |
|---|---|---|---|---|---|---|---|---|
| | B | Al | | | | In(?) | | Tl |
| | C | Si | Ti | | Zr | Sn | | Pb |
| | N | P | V | As | Nb | Sb | Ta | Bi |
| | O | S | Cr | Se | Mo | Te | W | |
| | F | Cl | Mn<br>Fe<br>Co<br>Ni | Br | Ru<br>Rh<br>Pd | I | Os<br>Ir<br>Pt | |
| Li | Na | K | Cu | Rb | Ag | Cs | Au | |
| Be | Mg | Ca | Zn | Sr | Cd | Ba | Hg | |

**System of elements, which Lothar Meyer has proposed 1870 (the figure was downloaded from the public web sites)**

| | No. | | No. | | No. | | No. | | No. | | No. | | No. | | No. |
|---|---|---|---|---|---|---|---|---|---|---|---|---|---|---|---|
| H | 1 | F | 8 | Cl | 15 | Co & Ni | 22 | Br | 29 | Pd | 36 | I | 42 | Pt & Ir | 50 |
| Li | 2 | Na | 9 | K | 16 | Cu | 23 | Rb | 30 | Ag | 37 | Cs | 44 | Os | 51 |
| G | 3 | Mg | 10 | Ca | 17 | Zn | 24 | Sr | 31 | Cd | 38 | Ba & V | 45 | Hg | 52 |
| Bo | 4 | Al | 11 | Cr | 19 | Y | 25 | Ce & La | 33 | U | 40 | Ta | 46 | Tl | 53 |
| C | 5 | Si | 12 | Ti | 18 | In | 26 | Zr | 32 | Sn | 39 | W | 47 | Pb | 54 |
| N | 6 | P | 13 | Mn | 20 | As | 27 | Di & Mo | 34 | Sb | 41 | Nb | 48 | Bi | 55 |
| O | 7 | S | 14 | Fe | 21 | Se | 28 | Ro & Ru | 35 | Te | 43 | Au | 49 | Th | 56 |

**System of elements proposed by John Alexander Newlands. (the facsimile was downloaded from the public web sites)**

During his regular lecture for students at the University of Sankt Petersburg, on 17 February 1869, Dmitri Ivanovič Mendeleev has outlined, from methodical reasons (!) the first version of the *periodic system of chemical elements* represented in the following figure.

| | | | | | |
|---|---|---|---|---|---|
| | | | Ti=50 | Zr=90 | ?[2]=180 |
| | | | V=51 | Nb=94 | Ta=182 |
| | | | Cr=52 | Mo=96 | W=186 |
| | | | Mn=55 | Rh=104,4[3] | Pt=197,4[4] |
| | | | Fe=56 | Ru=104,4 | Ir=198 |
| | | | Ni=Co=59 | Pd=106,6 | Os=199 |
| H=1[5] | | | Cu=63,4 | Ag=108 | Hg=200 |
| | Be=9,4 | Mg=24 | Zn=65,2 | Cd=112 | |
| | B=11 | Al=27,4 | ?[6]=68 | Ur=116[7] | Au=197? |
| | C=12 | Si=28 | ?[8]=70 | Sn=118 | |
| | N=14 | P=31 | As=75 | Sb=122 | Bi=210? |
| | O=16 | S=32 | Se=79,4 | Te=128? | |
| | F=19 | Cl=35,5 | Br=80 | J=127[9] | |
| Li=7 | Na=23 | K=39 | Rb=85,4 | Cs=133 | Tl=204 |
| | | Ca=40 | Sr=87,6 | Ba=137 | Pb=207 |
| | | ?[10]=45 | Ce=92[11] | | |
| | | ?Er=56 | La=94 | | |
| | | ?Yt=60 | Di=95 | | |
| | | ?In=75,6 | Th=118? | | |

**First outlines of the periodic system of elements of Dimitri Ivanofič Mendeleev**

**Dmitri Ivanovič Mendeleev (1834–1907)**

As a basis for the construction of the periodic system, Mendeleev has not used only the ratios of masses and valences. Rather, he has considered other chemical and physical properties. One of the most important property is the formation of seven different types of oxides of elements, such as $M_2O$, MO, $M_2O_3$, $MO_2$, $M_2O_5$, $MO_3$ and the group $MO_4$- in particular compounds (perchlorates, permanganates, etc.). In his revised table of elements Mendeleev has proposed the classification of elements in seven groups (columns in the Table) with the addition of the group of noble gasses, and in six periods (rows in the Table) [7].

The scientific value of the periodic system, or the possible periodic law, was in its ability to predict properties of elements, which were unknown at that time. Some places in the periodic table were remained empty waiting for the discovery of new elements. Mendeleev has called them *Eka-boron*, *Eka-aluminium*, and *Eka-silicon*. After the discovery of elements *scandium, gallium* and *germanium*, their basic properties were in high agreement with those predicted from their position in the periodic table. However, it must be mentioned that Mendeleev has anticipated the properties of some other unknown elements, but with wrong prediction of their properties [8]. Does it mean that Mendeleev periodic law should be, in accordance with Popper criteria, rejected from science? This represents an additional example for the discussion about the demarcation between good and wrong science (*vide infra*).

Fundamental principles of structural theory of microcosmos were emerged from chemical investigations, i.e. from the epistemology based on the method of analogy and classification, [9] much more than from the use of logical manipulations. The *building principle* in modern textbooks, by which the periodic table can be constructed from the electronic configurations of elements is often in a wrong way explained as the origin of the theory of periodicity of properties of elements. Representation of electronic structures of atoms of elements is very radical approximation [10]. The periodic system of elements is originally chemical concept, not the consequence of the discovery of electronic structures of atoms.

As I have argued, Dumas has used analogy as the method in the comparative classification of organic molecules and chemical elements. Now I wish to explain how the conceptual systems such as theory of types and periodic law can be used more generally for the clarification of metaphysical meaning of the notions *quality* and *quantity*. The representative models for this is the periodic system of elements. Its periods (rows) are organized according to the increase of relative atomic masses (in Mendeleev version). This principle is exclusively *quantitative*, because it is based on the numerical physical values. However, the groups (columns) are created by the type of compounds (for instance, oxides) that an element can form. If elements E form oxides $EO_2$, then all of them belong to the same group. Elements witin the same group are analogous, and they are classified not on the quantitative, but by using the *qualitative* criterion. Thus, the periodic table is a two-dimensional classification system in which abscise (period) is a measure for quantity, and the ordinate (the eight groups of elements) is the representation for quality.

Similarly, the system of type theory is also two-dimensional. The four structural types (types of HCl, $H_2O$, $NH_3$, and $CH_4$) exemplify four different *groups of quality* and the various compounds within the same structural type differ in quantity, i.e. their chemical composition represented by Berzelius chemical formula.

## 10.1 Conclusions

It is argued how the combination of methods of classifications with quantitative relations, as well as the use of analogy has led to the formulation of two most important concepts in chemistry, structural theory and periodic system of chemical elements. The structural theory and its representation with chemical formula and chemical nomenclature is investigated also as a model for the discussion about radical falsificationism proposed by Karl Popper. The case study about the structure of benzene molecule has shown that although the proposed chemical formula is logically inconsistent, it is still of a great scientific value. This problem in epistemology could be better comprised within the Quine-Duhem-Lakatos approach, which somehow includes epistemic values in science.

## References

1. Kekulé A (1857) Über die s. g. gepaarten Verbindungen und die Theorie der mehratomigen Radicale, Ann. der Chemie und Pharmacie, p 9
2. Berson JA (2003) Chemical Discovery and the Logicians' Program. Willey-VCH GmbH & Co., Weinheim
3. Ladenburg A (1869) Bemerkungen zur aromatischen. Berichte der Deutschen Chemischen Gesellschaft. 2:140–142
4. Curd M, Cover JA Eds. (1998) The Duhem thesis and the Quine thesis. In: Philosophy of science: the central issues. Norton, New York, pp 302–319
5. Lakatos I (1989) The methodology of scientific research programs. In: Worral J, Currie G (eds) Philosophical papers, vol 1. Cambridge University Press, NY
6. Dumas MJ (1857) Mémoire sur les équivalents des corps simples, Ann de chim et de phys, p 129
7. Mendeleef DI (1889) The periodic law of the chemical elements. J Chem Soc 55:634
8. Scerri ER, Woral J (2001) Studies in history and philosophy of science, Part A 32:407–452
9. Duhem P (1991) (1914) The aim and structure of physical theory. Princeton University Press, Princeton, NJ
10. Scerri ER (1997) the periodic table and the electron. Am Sci 85:546

# Models and Reality

# 11

*Symmetry is what we see at a glance.*

Blaise Pascal

**Abstract**

This chapter is dedicated to some problems about the use of models in the discovery of the spatial configurations of molecules and chirality. The appearance of the concept of chirality in the history of chemistry is discussed as a case study about the epistemic application of the methods of empirical induction and logical deduction. The appearance of physical chemistry, which is focused on the relationship between the observational parameters and the theoretical concepts developed within the structural theory, could be regarded as a new paradigm in science. By combination of physical and chemical concepts the chemical phenomena were more deeply explained in the frame of electron structure of atoms.

**Keywords**

Models · Chemical topology · Configuration · Stereochemistry · Induction · Deduction · Chemical atoms · Physical atoms · Chemical dualism · Gravimetry · Volumetry · Semiempirical approach · Chemical graph theory

As I have mentioned, the structural theory that has been developed from the type theory has introduced the *principle of constitution*, which determines the mode of the interlinking of atoms. Let me point out, that this constitutional representation does not include the real physical space. Constitution is a *topological category*. Topology is an abstract mathematical theory introduced in 17th century by **Gottfried Wilhelm von Leibnitz** (1646–1716), and later developed by **Leonhard**



H. Vančik, *Philosophy of Chemistry*, Integrated Science 2,
https://doi.org/10.1007/978-3-030-69224-7_11

**Euler** (1707–1783). This branch of mathematics is focused on the properties of geometrical objects, which are preserved under continuous deformations, such as stretching, bending or twisting. Chemical constitution comprehended as a mode of interlinking between atoms can serve as one of the best models for application of such an abstract mathematics. The topological analysis of chemical formulas is known in science as the graph theory.

Constitutional formulas are "plane-like", and, before all, conceptional: only the valence of elements, not the spatial structure is recognizable from them. The question about the reality of molecules as physical objects depends on the possibilities of experimentations. As Dalton has, by his transduction method with quantitative experiments, finally accepted atoms, the physicality of molecules as spatial constructs will in the same way emerge by using transduction.

**Joseph Achille LeBel** (1847–1930) was one of the closest collaborators of **Louis Pasteur** (1822–1895). Both were interested in the chemistry of salts of tartaric acids. Pasteur and LeBel were isolated ammonium-sodium tartarate in the form of the mixture of crystals (represented on the following figure), which are identical by shape, but which are related to each other similarly as left and right hand [1].

**Left-hand, and right-hand forms of ammonium-sodium tartarate (The figure is downloaded from the public web site)**

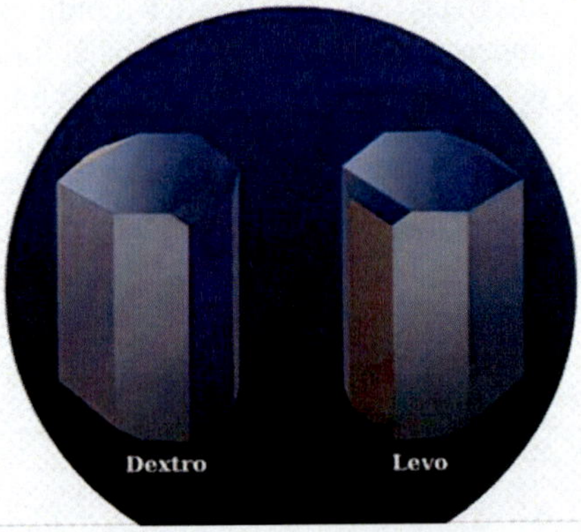


As there is no spatial re-orientation by which it is possible to overlap left and right hand, there is no way to overlap these left and right crystals by their reorientation in the space. Because the LeBel crystals have shown the same geometrical property as hands, they are called to be *chiral*, i.e. hand-like (Greek, χ, ε, ι, ρ, hand). Perhaps, it is not necessary to mention that geometrical solids which can be overlapped with their mirror image are not chiral, for instance the ball or the cube.

Chirality of crystals of tartarate has two possible explanations: either the molecules are packed in the crystal lattice in two mirroring ways, or the molecules are intrinsically chiral. In the second case, if molecules are intrinsically chiral, then they must have a defined spatial structure. LeBel has proposed the solution in which chirality of a molecule is possible only if the shape of molecule is in the form of tetrahedron with carbon atom in the middle.

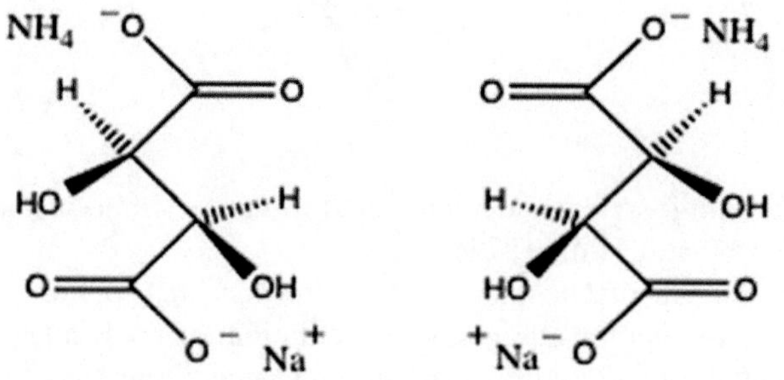


LeBell has proposed this solution after the experiment in which the solution of left crystals is different from the solution of right crystals in its property to rotate the plane of the polarized light in opposite direction but for the same angle. Thus, the solution of left and the solution of right crystals are also "chiral". Since in solution exist only molecules, not crystals, the conclusion that molecules as entities are chiral, is obvious. Polarized light that propagates only in one plane has been earlier discovered by physicist **Jean Baptist Biot** (1774 –1862).

Independently, **Jacobus Henricus van't Hoff** (1852–1911) has studied the possibility of representation of isomers by using constitutive formulas within the frame of the actual structural theory. For example, why dichloromethane can be represented with two structural formula, although it is the same compound?

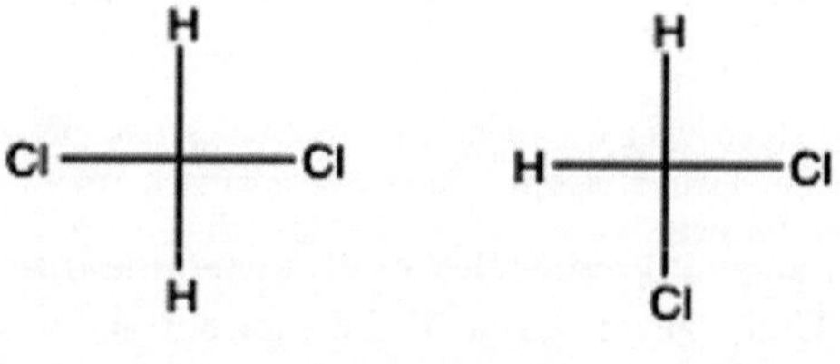


The possible answer is in the spatial structure of the molecule in which carbon atom is in centre of tetrahedron. This presumption can be tested by varying atoms linked to this central carbon, and testing whether the molecule becomes chiral or not. The presence or the absence of chirality is, thus, the decisive test for the validity of van't Hoff theory about the tetrahedral shape of molecules: chiral are only the molecules in which all the four atoms or groups which appear on the vertices of tetrahedron are different. The conclusion is clear, molecules possess spatial structure based on tetrahedrons with carbon atom in the middle. It follows that dichloromethane is described with only one formula which represents its spatial structure:

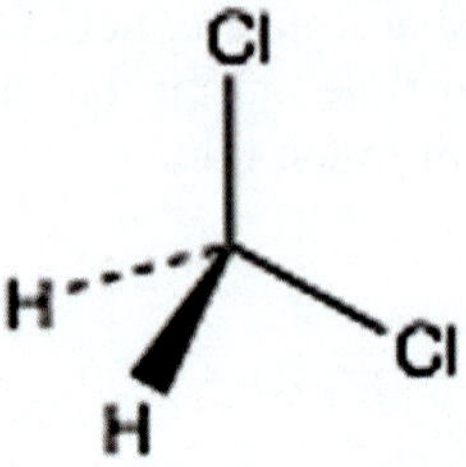


Such reasoning is clearly demonstrated in the facsimile of the original work of van't Hoff [2] (van't Hoff, 1874).

The new branch of chemistry, the *stereochemistry*, is born, and the structural theory based on composition, and constitution of molecules is extended with additional category—the *configuration*, the spatial arrangement of atoms. While in organic chemistry the basic spatial unit is tetrahedron, in inorganic chemistry this is octahedron, as **Alfred Werner** (1866–1919) has discovered.

LeBel and vant' Hoff have revealed the physical reality of molecules, their spatial shape, independently on each other not only in sense that they did not collaborated in this discovery, but also in their use of the opposite epistemologies [3]. LeBel is positivist because he has started with observation of chiral crystals and rotation of the polarized light as empirical base which led him inductively to the discovery of tetrahedral structure of organic molecules. LeBel is empiricist, and his method is inductive, i.e. from observation to theory.

On the other hand, van't Hoff is deductivist, he starts with the conjecture about the possibility that organic molecules could have the tetrahedral shape, and then he has tested this conjecture by consideration of the consequences of molecular constitution on the chirality [3]. Karl Popper would be happy with this van't Hoff's approach.

> The same theory about spatial structure of molecules has emerged from two opposite epistemologies, the inductive, and the deductive approach.

Discovery of atoms and molecules has initiated the discussion about the nature of these entities. Until the proposal of the concept about the spatial structures of molecules, the scientists and philosophers were divided in their opinions. Some of them have regarded atoms and molecules only as a conceptual model by which chemical quantitative laws can be explained («chemical atoms»), and others have argued that atoms are real particles («physical atoms»). Even August Kekulé has initially used atoms as abstract concept [4]. However, in the further development of science, the idea about chemical atoms has gradually vanished.

However, it must be pointed out that the idea about *conceptual particles*, as well as the idea about *conceptual structures* is in chemistry still unavoidable. In their everydays practice, chemists manipulate with the abstract spatial models of molecules represented in formulas, either for the explanation of reaction mechanisms, or for designing syntheses of novel materials. From the physical point of view, the rigid molecular structures with the determined spatial configurations of atoms do

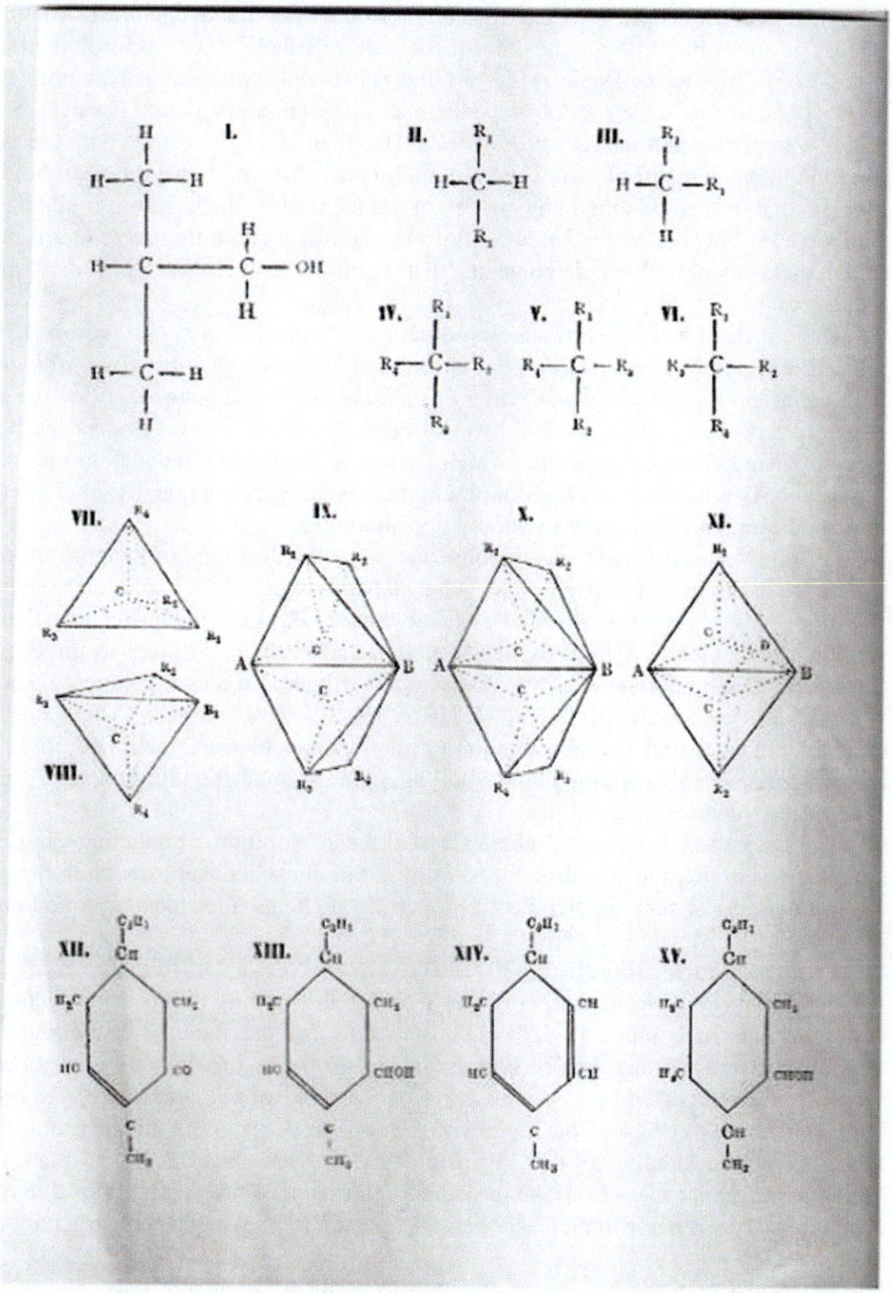


**Facsimile of the van't Hoff's paper from 1874**

not exist as a classical "Newtonian" reality. On the other hand, chemical structural theory is ufeasible without the introduction of relatively drastic approximations used from physics: molecule possesses the "chemical structure" only within the assumption of so called Born-Oppenheimer approximation (**Max Born**, 1882–1970, **Robert Oppenheimer**, 1904–1967). This heuristic axiom starts with the fact that the movements of electrons are so much faster than the movements of atomic nuclei, that we could build the model of molecule in which atomic nuclei are immovable. This conceptualism of entities in chemistry is not limited to atoms and molecules. Soon we will demonstrate that electrons can also be "physical" and "chemical".

The conclusion that follows is important: *the same category, for instance the structure, or electrons, could be considered within different epistemologies depending on the level of complexity in which the entities in consideration appear.* Emergent properties on chemical level of complexity can be analysed in the best way by using chemical concepts, which cannot be completely reduced to physical concepts. As it has been analysed in the first chapter, every complexity level can be well represented exclusively by its own epistemology.

LeBel's key experiment was the observation of the declining of the plane of the polarized light in left of right side, what depends whether the solution contains "left-handed" [*l*] or "right-handed" [*d*] molecules. Physical parameter, the declination of the polarized light (optical rotation) LeBel has associated to molecular structure. This important step in chemistry has triggered the establishment of new branch of chemistry, the *physical chemistry* (or, *chemical physics*). The scope of this field of science is the investigation of relationships between measured physical parameters such as melting points, boiling points, refractive index, density, and chemical structures of molecules.

The basic task of physical chemistry has been ambitious: predicting physical parameters of particular substance starting from the chemical structural theory. Some attempts of such predictions of properties of chemical elements can be found already in the Mendeleev periodic law.

**François-Marie Raoult** (1830–1901) was professor at the University of Grenoble who has decided to correlate melting and boiling points with chemical compositions of compounds. All chemists know for the Raoult's law about the partial pressures in mixtures of liquids: *partial pressure of vapour of every component of the ideal mixture of liquids is equal to the vapour pressure of the pure component multiplied with the molar fraction of this liquid.* Without entering in the details, we can simply say that, by using this law, it is possible to calculate the relative molecular mass from the measured depression of melting point of different mixtures. The relative molecular mass is related to the molecular structure, its composition.

Physical chemistry has, more or less successfully, found not only a series of correlations of physical measurements with molecular structures, but also the laws of the rates of chemical reactions. Van't Hoff has systematically measured the rates of chemical reactions establishing in this way *chemical kinetics* as a new branch of chemistry. Van't Hoff has also defined a notion of molecularity of reactions, i.e.

how many of different sorts of molecules determines the rate of the chemical reaction. His introduction of the terms such as *unimolecular*, and *bimolecular reaction*, was of far-reaching importance for the future development of the theory of reaction mechanisms.

**Wilhelm Friedrich Philipp Pfeffer** (1845–1920) has systematically measured osmotic pressure, the force that appear in systems in which two solutions of different concentrations are separated with the semipermeable membrane. The solution with higher concentration induces pressure on the membrane in direction to the solution with lower concentration. Stimulated with the Pfeffer's measurements, van't Hoff has demonstrated that the ratio between concentration and the osmotic pressure is analogous to the ratio of the volume and pressure of ideal gas. Thus, his law of osmotic pressure is analogous to the Boyle-Mariott law of ideal gas.

**Harmon Northrop Morse** (1848–1920) has extended van't Hoff's law for all, organic and inorganic solutions. However, this extension has in the equation an additional factor *i* that was an empirical parameter. Since it was not possible to calculate the factor *i*, the Morse equation is *semiemprical*, i.e. its solution depends on the parameters provided exclusively from the observations. *Semiempiria*, the improvement of equations by addition of some parameters obtained by experience, became in this way part of epistemology, and this method will play very important role in the development of methods of calculations of molecular structures. On the other hand, the theories that are based on the "fundamental truth", i.e. on the heuristic statements, so called *ab initio theories*, are in reality out of use because their calculi are highly complex. Science, and especially chemistry, has developed its epistemology that is basically *semiempiric*.

Although, at the first sight, is seems that the semiempirical character of science is a signal for its weakness, it should be pointed out that the introduction of experienced parameters in calculations represents a germ of new theoretical extensions.

Discussion about the Morse factor *i* has initiated new discovery for which the most of the credit should be given to **Svante August Arrhenius** (1859–1927). In the year 1887 Arrhenius has created the theory about the electrolytic dissociation: acids, bases, and salts are in water solution partially separated in anions and cations. From the ratio of concentrations of ions and nondissociated molecules (*the extent of dissociation*) Arrhenius has calculated the factor *i*. Thus, semiempirical addition to theory has initiated its extension to new theory in which this semiempirical factor acquires unequivocal physical meaning. **This interplay between introducing and vanishing of semiempirical entities is one of central forces in the development of science**.

Arrhenius idea about the equilibrium of dissociated and non-dissociated molecules was generalized in the *law of dilution* discovered by **Friedrich Wilhelm Ostwald** (1853–1932) , the Nobel laurate for the year 1909, and in the *law of chemical equilibrium* formulated by **Peter Waage** (1833–1900) and **Cato Maximilian Guldberg** (1836–1902).

The discovery that the disturbed system always has a tendency to be equilibrated, which is in chemistry known as *Le Châtelier principle* (**Henry Louis Le**

**Châtelier**, 1850–1936), is of general importance in science, and its origin was in the studies of chemical equilibrium.

Arrhenius theory about electrolytic dissociation is direct consequence of one of the crucial discoveries in science. **Alessandro Giuseppe Antonio Anastasio Volta** (1745–1827), the Italian chemist and physicist, has in 1799 and 1800 performed experiments, which resulted with the discovery of electrical current. The first source of electrical current Volta has constructed from copper and zinc plates that in direct contact induce electrical potential. Before this discovery—invention, the electricity has been regarded mostly as the appearance characteristic for living beings. Such a "vitalistic theory" has origin in investigations of **Luigi Aloisio Galvani** (1737–1798), the physiologist who has observed electrical effects in frogs. Nonphysiological source of electrical current invented by Volta has also staggered the vitalistic point of view, the concept that has finally been abandoned after the work of Friedrich Wöhler, as it was discussed in detail in previous lectures.

Consequences of this Volta's discovery were enormous. Electrical current became one of the most important tools in experimental chemistry and physics. By immersion of electrodes in solution, chemistry was oriented to method of electrolysis. The Lavoisier's theory about water as a compound consisting from hydrogen and oxygen has been supported by electrolysis.

In chemical epistemology, electrochemical experiments have two consequences: the rejection of *vitalism*, and establishing *chemical dualism*! Systematic experimentation with electrolysis has shown that the substances can be divided in two classes. The substances that in electrolysis are developed at the positive electrode, the **anode**, belong to the first group, and the substances that appear on the negative electrode, the **cathode**, belong to the second group. For instance, metals always collect on cathode, and the molecules which have oxygen atom of another non-metal appear on anode. But electrolysis can be applied not only to solutions, but also to melts of salts. As I have already discussed, thanks to the invention of the platinum laboratory equipment, Humphry Davy has electrolysed melted salts and isolated for the first time potassium, sodium calcium, etc.

If decomposition of substance yields two components which are electrically opposite, then chemical affinity, the force that keeps atoms together, must be electrical by nature. If chemical affinity is electrical by its nature, then only different atoms could be bound. Identical atoms cannot form chemical bond; this is the basic idea of the chemical dualism. Hydrogen cannot bind hydrogen, or, oxygen cannot bind oxygen, but hydrogen can be bound with oxygen to form water. As it has been mentioned, the most prominent representative of chemical dualism was Jöns Jacob Berzelius, one of the most meritorious chemist in the development of structural theory.

Although dualism was an advancement in science because it has linked chemical with electrical appearances, in spite to that, it became an obstacle for the development of structural theory of complex organic molecules. Discovery of molecules in which carbon atoms are bound to each other (Kekulé, Couper) has not been the only overcoming of dualistic approach.

Both, the discovery of simple molecules as well as Lavoisier's scientific revolution, were consequences of the systematic studies of chemistry of gases. Thanks to the accurate measuring it has been possible to investigate in detail the relationship between mass and volume of various gases. Besides the *gravimetry* that comprise the measurements of mass ratios (laws of Dalton and Proust), the studies were extended with the *volumetry*, the systematic investigation of ratios of volumes of gases. **William Prout** (1785–1850) has found that the ratios of masses of equal volumes of different gases (under the *ceteris paribus* conditions of constant pressure and temperature) are whole numbers. This observation Prout has compared with the fact that relative atomic masses of elements are in principle the integer products of the relative atomic mass of hydrogen. For instance, one volume unit of chlorine is 36 times heavier that the same volume of hydrogen, or, the equal volume of oxygen is 16 times heavier than hydrogen. From these mass ratios for equal volumes Prout in 1820 derives revolutionary hypothesis: there is only one chemical element, hydrogen, and all the other elements, i.e. their atoms, are composed from different numbers of hydrogen atoms. Revolutionarity of this idea is far-reaching because it implies that atoms are composed from smaller particles—the hydrogen atoms, which Prout has considered as real elemental particles, giving them the name *protyle* (the term composed from Greek words *πρωτη υλη*). **Ernest Rutherford** (1871–1937), the physicist who has discovered the structure of atoms in 20th century, gave to this fundamental component of atom the name *proton*, in honour to William Prout.

By his hypothesis Prout has not only announced the concept about the structure of atom, he has also established the basis of volumetry, the method from which one of the most important scientific categories has emerged—the *amount of substance* (say, the number of particles). If the volume of chlorine is for 36 times heavier than the same volume of hydrogen, and if atoms of chlorine are merely aggregates of atoms of hydrogen, then **equal volumes of gases under the *ceteris paribus* conditions must be composed from the equal number of particles!** However, history of science has this law attributed to the famous Italian chemist. In its final form this law has been defined by **Lorenzo Romano Amedeo Carlo Avogadro di Quaregna e di Cerreto** (1776–1856) and all of us know it as Avogadro law [5]. This law is stimulating because it opens the question about the number of particles that appear in the referent volume—about *their amount*—the **mole**. The unit for the amount of substance, the new parameter in natural science, is called **mol**—the number of molecules in standard volume, and its numerical value, i.e. the number of particles in 1mol, is known as *Avogadro's* constant, although it has been determined many decades after Avogadro. The approximate value of Avogadro constant, $2.686\ 7805(24) \times 10^{25}$ molecules per cubic meter at 0 °C, and 1 atmosphere pressure has for the first time determined by **Johann Josef Loschmidt** (1821–1895), who has applied the kinetic theory of gases. In literature, this number is known as *Loschmidt number*.

The idea about the constant number of particles in given volume (under the *ceteris paribus* conditions) has brought Avogadro and his collaborator Cannizzaro (**Stanislao Cannizzaro**, 1826–1910) to the final *discovery of molecules consisting* *from identical atoms*, such as hydrogen with hydrogen ($H_2$), or, oxygen with oxygen ($O_2$). Berzelius dualism, the hypothesis that only non-identical atoms can combine in molecules, was with this discovery finally rejected. That was one of the most important steps in the development of structural theory.

Avogadro's and Canizzaro's discovery of two-atomic molecules has been the direct consequence of the fundamental *law of volumetry* formulated by **Joseph Louis Gay-Lussac** (1778–1850) in 1808: **in chemical reaction gases combine in constant ratios of volumes**. This is the same year when Dalton has published the law about multiple ratios of masses in his *A New System of Chemical Philosophy*, i.e. the basic law of *gravimetry*!

However, Dalton's law of gravimetry, and Gay-Lussac's law of volumetry were apparently in contradiction. While Dalton's law has confirmed the Lavoisier's discovery about the conservation of masses before and after chemical reaction, what means that the sum of masses of substances before reaction is equal to the sum of masses of substances after the reaction, from the Gay-Lussac's law does not follow that the volumes are the same before and after the chemical process. For example, Gay-Lussac and **Friedrich Wilhelm Heinrich Alexander von Humboldt** (1769–1859), who is better known by his ideas about the basic principles of modern universities, have already in the year 1805 discovered that *two volumes of hydrogen* and *one volume of oxygen* yield *two volumes of water*. Thus, since before the reaction there are three volumes (one of oxygen, and two of hydrogen), and after the reaction we have only two volumes of water vapour, the total volume before and after the reaction is not the same.

The nonexistence of the law of conservation of volume has stimulated Cannizzaro and Avogadro to propose that hydrogen, oxygen and other gasses exist as two-atomic molecules $H_2$, $O_2$, etc. Berzelius representation of the reaction must to be reshaped in chemical equations, such as we use today:

$$2H_2 + O_2 \rightarrow 2H_2O$$

Law of conservation of masses is conserved by the discovery of the two-atom molecules, because from the upper equation it is clear that the number of atoms is the same before and after the reaction (in summary, 4 hydrogen atoms and 2 oxygens on both sides of the equation). Let me mention that Gay-Lussac has also formulated an additional physical law about the correlation of pressure or volume with temperature. This law plays the important role in classical thermodynamics.

Accumulated discoveries of new phenomena such as atoms, molecules, amount of substance, atomic masses molecular masses, etc., were required a new kind of the general systematization, a sort of *convention* between chemists. Such convention, which was especially supported by August Kekulé, and which comprises definitions of atoms, molecules, and the values of relative atomic masses of all known elements, was the result of the first world congress of chemists organized in Karlshuhe (Germany) from 3 till 5 September 1860. The **Karlsruhe congress** has been the beginning of standardization in chemistry, necessary for the further development of science and its applications. The agreed values of atomic masses, nearly 1 for hydrogen, 12 for carbon, and 16 for oxygen, were the immediate result of the Cannizzaro's idea about the existence of two-atomic molecules. Congress in Karlsruhe was the germ for the later establishment of the International Union of Pure and Applied Chemistry (I.U.P.A.C.).

Volta's discovery of electrical current and electrolysis has been extended in new branch of chemistry, the *electrochemistry*, which became a quantitative discipline after the work of **Michael Faraday** (1791–1867), who found the correlation between the mass of the substance produced by electrolysis with the amount of the charge used for this reaction. In this way, the value *amount*, which is expressed in units called **mol**, could be used not only for the amount of substance, but also for the amount of charge. By Faraday, 1 mol of charge corresponds to 96500 C (C, coulomb, is the unit for charge), the number that is known as *Faraday constant*. The remained open question was the nature of the carrier of charge. The answer has emerged at the end of 19th century when **Sir Joseph John Thomson** (1856–1940) has discovered *electron*. Thomson considers electron as a part of atom, which is responsible for *chemical affinity*—the binding of atoms in molecules. Electrons of atoms that bind each other form pairs called *chemical bond*, the basic phenomenon that determines molecular structure [6, 7]. This was the beginning of the concept of *chemical bond as an electron pair*, the concept that will finally reshape the *classical structural theory*.

In 1910, **Robert A. Millikan** (1868 –1953) has succeeded to determine experimentally the charge of electron that is equal to $1{,}602 \times 10^{-19}$ C. This measurement has enabled the accurate calculation of Avogadro constant ($N_A$) by dividing Faraday constant with the charge of electron: $(96400\ \mathrm{C})/(1{,}59 \times 10^{-19}\ \mathrm{C}) = 6{,}025 \times 10^{23} = N_A$. Numerical determination of Avogadro's constant was the crucial scientific achievement because it has opened the door for the immediate study of microcosmos by using exact experimental methods. Measured parameters obtained by the observation of systems with particular *amount of substance* could be, by using Avogadro constant, easily recalculated in the properties of single molecule. Dalton's transductive approach in connecting abstract concept with the measured ratios of masses now seems visionary. Thus, it must be pointed out that Dalton did not only discovered atoms, but he also made s step forward in the development of epistemology of natural sciences.

The method of transduction has been proliferated in science after the construction of new scientific instruments. **Robert Wilhelm Eberhard Bunsen** (1811–1899) and **Gustav Robert Kirchhoff** (1824–1887) have revealed their first studies about *spectral analysis*, the method that is today known as *spectroscopy*, the branch of science that connects observable characteristics of spectra with chemical composition and structure. In the beginning, the spectra analysed within the visible region were used for the characterization of metals. Very interesting was the fact that the same spectrum has been obtained either by observation of stars, or by observation of hot metal vapours in the flame. The spectrum is, thus, directly connected to the identity of metal—the chemical element. However, for the linkage of abstract theory of molecular structure with observations, the additional methods were necessary. Discovery of X-rays by **Wilhelm Conrad Röntgen** (1845–1923) has opened the possibility of the immediate measurement and description of configurations of molecules. Roentgen structural analysis through the diffraction of X-rays, developed by **Max von Laue** (1879–1960), **Sir William Henry Bragg** (1862–1942) and **Sir William Lawrence Bragg** (1890–1971) has the idea about the spatial structure of molecule for the first time converted into realistic picture: the molecules have really a defined configuration of atoms in the space.

However, the relationship between empirical experience and its representation, in chemistry it is the relationship between chemical formula and measurable properties of substance, is full of puzzles. As I have already announced, in the second half of 20th century the new concept, the *graph theory* has been developed. The scope of this theory is to find the relation between chemical formula viewed as a graph, the drawing, with the properties of chemical compounds represented by it [7, 8]. *Chemical formulas are not only the icons, rather they are bearers of information which should still be discovered.*

Chemical graph theory is focused on the problem of chemical formulas. The triad, *contents*, *language*, and *script* is the foundation of chemical epistemology, where contents are a real chemical substance and its change, language is chemical nomenclature, and script are chemical formulas. Formulas as chemical script are not merely labels for corresponding substances, or for their molecular constitution or configuration. Pioneers of chemical graph theory, **Alexandru Balaban**, **Frank Harary**, **Haruo Hosoya**, **Ante Graovac**, **Ivan Gutman**, **Milan Randić**, **Nenad Trinajstić**, **Harry Wiener**, have opened the questions about the metaphysics of chemical script. As it is represented in following figure, the graph theory is in principle a sort of *meta-representation*: while molecular formula is the representation of molecular structure, chemical graph is the representation of molecular formula, i.e. it is the representation of representation of the structure. Algebraic *matrix of neighbourhood* (the numbers show in which neighbourhood relations are vertices in the graph) is the further metarepresentational step, the meta-meta-representation of molecular structure. The result of this approach is surprizing: by the numerical manipulation with meta-meta-representations it is possible to predict properties of substances merely by looking in their molecular graphs!

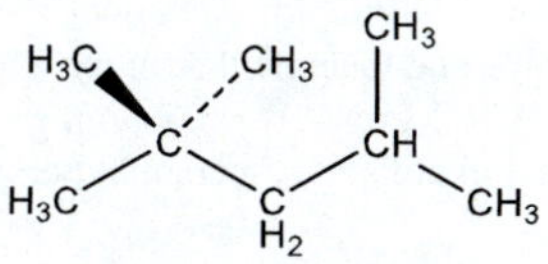

Molecular formula
Representation of molecule

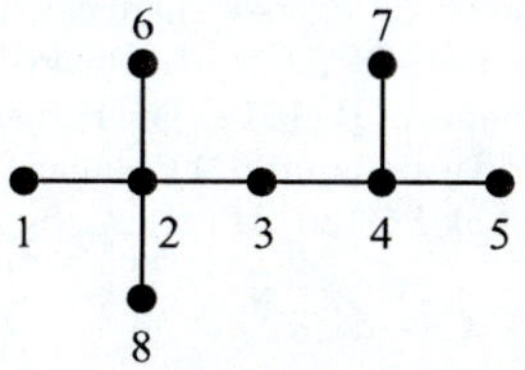

Molecular graph
Representation of molecular formula

| | | | | | | | |
|---|---|---|---|---|---|---|---|
| 0 | 1 | 2 | 3 | 4 | 2 | 2 | 4 |
| 1 | 0 | 1 | 2 | 3 | 1 | 1 | 3 |
| 2 | 1 | 0 | 1 | 2 | 2 | 2 | 2 |
| 3 | 2 | 1 | 0 | 1 | 3 | 3 | 1 |
| 4 | 3 | 2 | 1 | 0 | 4 | 4 | 2 |
| 2 | 1 | 2 | 3 | 4 | 0 | 2 | 4 |
| 2 | 1 | 2 | 3 | 4 | 2 | 0 | 4 |
| 4 | 3 | 2 | 1 | 2 | 4 | 4 | 0 |

Algebraic representation of molecular graph

## 11.1 Conclusions

It is argued how both the epistemic methods, the method of induction and the method of deduction, respectively, can independently contribute in the discovering of new phenomena, in our example, the phenomena of chirality and spatial structure of molecule.

New branch of chemistry, the physical chemistry has extended our knowledge not only in the quantifications of relationship between the empirical observables and chemical concepts, but also by providing new models for the representation of molecular structure.

## References


1. Tobe J (2003) Mendeleev Commun 13:93–94
2. van't Hoff JH (1874) Voorstel tot uitbreiding der tegenwoording in de scheikunde gebruikte structur-formules, Utrecht, J. Greeven. (English transl. Benfey OT (1974), Dover Pub. Inc. N. Y., p 151–160)
3. Cintas P (2002) Found Chem 4:149–161

4. Rocke AJ (1984) Chemical Atomism from Dalton to Cannizzaro, Columbus.
5. Murrell JN (2001) Avogadro and his constant. Helv Chim Acta 84:1314
6. Thomson JJ (1914) The Forces Between Atoms and Chemical Affinity, Phil. Mag. S.6. 27: 757.
7. Trinajstić N (1992) Chemical graph theory. CRC Press, Boca Raton, Florida
8. Basak SC (2013) Philosophy of mathematical chemistry: a personal perspective. HYLE 19:3–17

# Limits of Structural Theory 12

*What we observe is not nature itself, but nature exposed to our method of questioning.*

Werner Heissenberg

**Abstract**

Electronic structure of atoms and molecules, as well as the consequences of this discovery on the general ways of thinking in science is discussed. It is demonstrated how the knowledge about the limitations of the Lewis structural theory has appeared in the moment when the structural theory has been applied in the theory about reaction mechanisms.

**Keywords**

Electrons · Identity · Valence · Bohr's theory of atom · Bohr's theory of molecule · Electronic structure of molecule · Magneton · Lewis electron pairs · Theory of reaction mechanisms · Quantum mechanics · Nonclassical molecular structures · Philosophical generalization of the uncertainty principle

After Thomson's discovery of electrons, and after the presumption that pairing of these particles can be the real nature of chemical affinity, say, chemical bond, the binding lines in the Crum Brown formulas acquire physical meaning. **Gilbert Newton Lewis** (1875–1946) has established, starting from this idea, a new form of the structural theory, which is based on two fundamental presumptions [1]. In the first presumption, chemical bond is a pair of electrons, and it is in chemical formula represented by the binding line. The second presumption is for science much more far-reaching: *electrons do not have identity*! In the accordance with my

 
H. Vančik, *Philosophy of Chemistry*, Integrated Science 2,
https://doi.org/10.1007/978-3-030-69224-7_12

classification of the category of identity explained in previous lectures, though, electrons possess the *primary identity*, the identity of substance.

To make his concept of chemical bond as a pair of electrons persuasive, Lewis has proposed the basic heuristic fundamental model about the distribution of electrons in atoms.

The independent discovery that has been drawn from the classical theory of valence, which is known as an *octet rule*, helped Lewis to complete his theory about electronic structure of atoms and molecules. German chemist **Richard Wilhelm Heinrich Abegg** (1869–1910) has referred to the possibility that the valence is composed from two components: the valence, and the counter-valence, respectively [2]. As the electrochemical dualistic theories have suggested, the chemical affinity (chemical bond), since its nature is "electrical", must include a sort of "valence polarity". Abegg's *valences* and *counter-valences* have the opposite charge. For instance, chlorine can have the valence +7 in perchlorates, and the counter-valence −1 in chlorides. By connecting the concept of the valences and counter-valences with the periodic law, Abegg has constructed the tabular representation from which it follows that *for all the elements, the sum of valence and counter-valence is equal to 8*, the idea known as the *octet rule*. The following table shows the relationships of valences and contravalences for some elements of the main period of periodic system. It can be clearly recognized, that the octet rule imagined in this form is idealized. While in the case of chlorine and sulphur it is possible to find their compounds with both, the valences and the counter-valences, it is difficult even to imagine the compound of sodium with the counter-valence −7, or the compound of magnesia with the counter-valence −6.

Within the Popperian analysis of concepts of theories as the refutable phenomena, the Abegg rule cannot pass the test. However, we have already demonstrated in the case of the Kekulé's benzene, that the Popperian criterion should be used with care. Although the octet rule doesn't have an unequivocal experimental support, its conceptual value was out of question because it has stimulated further development of chemical theory.

| Group i the Periodic system | 1. Na | 2. Mg | 3. Al | 4. C | 5. P | 6. S | 7. Cl | 8. Ne |
|---|---|---|---|---|---|---|---|---|
| Valence | +1 | +2 | +3 | +4 | +5 | +6 | +7 | +8 |
| Contra-valence | −7 | −6 | −5 | −4 | −3 | −2 | −1 | 0 |

The octet rule, although provisory because it cannot be completely supported by examples, has inspired Lewis in his concept that the maximal number of electrons responsible for the formation of chemical bond corresponds to the Abegg's number 8.

Lewis theory of chemical bond, or, better to say, his representation of chemical bond, is of such importance for science and for philosophy of science, that I will deserve more attention to it. This chemical theory has appeared three years after the *Bohr's theory of atom* [3] (**Niels Bohr**, 1885–1962), and it get ahead the new quantum mechanics proposed by **Werner Heisenberg** (1901–1976), in 1925, its second interpretation by **Erwin Schrödinger** (1887.–1961.) in 1926, and the first quantum-chemical calculation of the molecule published by **Walter Heinrich Heitler** (1904–1981) and **Fritz Wolfgang London** (1900–1954) [4]. I will argue that the Lewis theory about the electronic structure of molecules, which has **its origin in chemical reasoning**, is conceptually and metaphysically analogous to physical quantum mechanics that has appeared within the rigorous physical laws of motion.

As I have demonstrated in previous lectures, chemical scientific methodology is based on classifications. Creation of the structural theory and periodic system of elements are the illustrative examples of the effectiveness of this epistemology in the investigation of the microcosm. Lewis uses the same methodology in building his theory about the chemical bond starting with chemical classifications: (i) dividing molecules in polar and nonpolar, and (ii) using the Abegg's octet rule. Pierre Duhem has argued that the method of classifications is in the relation with a sort of "natural classification", and that science oriented to the classification method led science to form the deeper insight in the nature, than the science that forms only the theories as its final product [5, 6]. Thus, science has to be oriented to classification. Perhaps, the development of chemistry is an additional argument for the thesis of Duhem.

For the beginning, I wish to focus on the theories of chemical bond, which have preceded the Lewis concept. One of the oldest ideas, which in principle talks about the nature of bonding of atoms is the theory about the atoms—material points formulated by Rugjer Josip Bošković. Since this theory has been analysed in the fifth lecture, and because this theory does not include the knowledge about electrons, I will not discuss it here.

J. J. Thomson, who discovered electrons, has proposed that these negatively charged particles should play the central role in the explanation of the chemical affinity, i.e. the chemical bond. In his work *The Forces between Atoms and Chemical Affinity*, [7] published in Philosophical Magazine, Thomson has established some of the basic concepts about chemical bond: (1) for binding of atoms in molecules the key role have electrons (Thomson has called them *corpuscules*); (2) electrons in atoms are separated in two "layers" of which only the so called *outer electrons* contribute to the formation of chemical bond, and the inner electrons remain stable, and bound to the atom; (3) maximal number of outer ("chemical") electrons cannot be higher than 8, and it depends on the position of the element in one of the 8 groups of the Mendeleev periodic system of elements; (4) chemical bond is formed by pairing of two electrons in such a way that one of bonding atoms donates its electron to another atom.

This shift of the electron from one to another atom Thomson describes as a "direction" of the mobile electron from the surface of the first atom to the centre of the second atom in the molecule as a consequence of the action of the special *tube of force* (the term used by Thomson). For instance, carbon atoms in the ethane molecule are bound in a way that the two atoms direct electrons to each other. While in the case of ethylene every atom directs two electrons, in acetylene there are three directed electrons. This description is shown in the picture taken from the original Thomson's publication.

**Facsimile from the Thomson's paper about the nature of chemical bond**

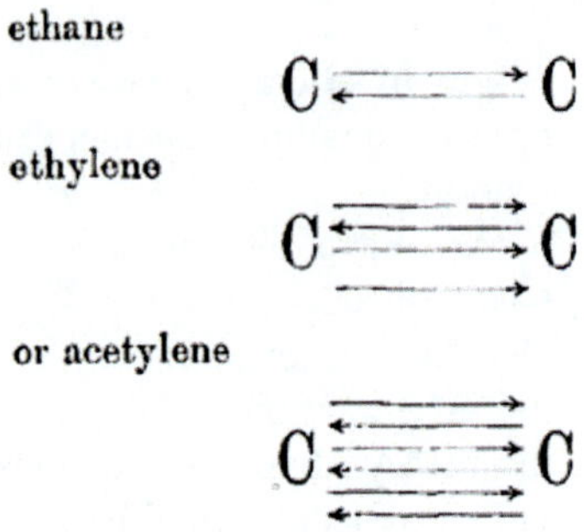


In spite to its epistemic arbitrarity, Thomson's theory could, in principe, solve the Kekulé's contradiction about the structure of benzene. In this molecule, every carbon atom directs one electron to every of its neighbours, and one electron in the circle. Consequently, the molecule must be six-fold symmetric, as it has later been experimentally shown to be true. The figure shows the representation of the structure of the benzene molecule, demonstrated in the Thomson's original work.

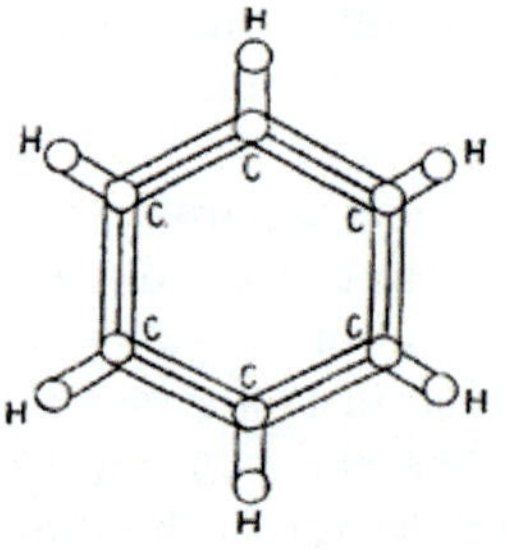


**Facsimile from Thomson's work about the nature of chemical bond in benzene molecule**

This is another demonstration how the scientific truth could origin from the incomplete theory, which as such could not pass Popper's test. However, Thomson has also proposed an additional type of bonding, the model of atom with the electropositive, and electronegative valences, similarly to the Abegg valences and counter-valences. For the instance, chlorine atom with seven outer electrons is by Thompson described with the ring consisting from eight negative, and one positive *corpuscule* (the term that Thomson has used for electron). Such a chlorine atom is capable to accept the eighth electron from another atom (for instance, the hydrogen atom). Such sort of interatomic bond is characterized by high polarity, and the corresponding molecules possess electrical dipole (Thomson has called it *electric doublet*). Evidently, the discovery of the polarity of molecules, as well as the discovery of dipole moment should be attributed to Thomson.

This discussion clearly shows that subdivision of chemical affinities (chemical bonds) in the polar, and nonpolar ones, respectively, has its origin in chemical classification. Thus, Thomson theory has to be regarded as a consequence of the research within the frame of chemical epistemology.

In physics, the atomic and molecular structure was analysed in the accordance with physical epistemology, in that time based on classical mechanics and electrodynamics. In this picture, the atom is a sort of microcosmic planetary system where the positive nucleus is surrounded by electrons in their orbits. This representation of atom is invented by Niels Bohr in his papers in *Philosophical Magazine*, published during the year 1913. In his Newtonian image of atom, Bohr has successfully calculated the electronic spectra of hydrogen atom, but nothing more. The calculated spectra for atoms of other elements did not agree with observations.

If, by Bohr, atom is a mechanical system with proton in the centre and the orbiting electron, then the hydrogen molecule should also be a sort of mechanical system [3]. Bohr has proposed that $H_2$ molecule consists from two atomic nuclei (protons) fixed on particular distance, and the two electrons orbiting in the plane perpendicular to the interatomic line:

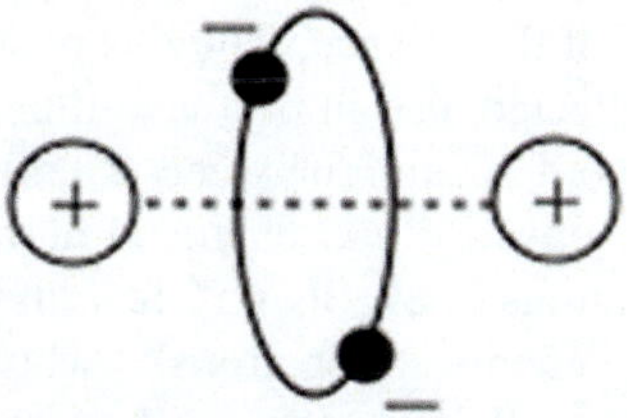

**Model of the structure of hydrogen molecule proposed by Niels Bohr**

By using such dynamical model, Bohr has calculated the energy required for the dissociation of the $H_2$ molecule in two hydrogen atoms (*the energy of chemical bond*) to be 60 kcal/mol (250,8 kJ/mol). The value accepted today (104 kcal/mol, 434,72 kJ/mol) markedly declines from this prediction. From these data, it is clear that the classical-mechanical Bohr's model of molecule cannot be applied for structures consisting from more atoms. It seems that classical physics has reached its "Gödelian limit". The fundamental change of scientific paradigm has appeared on the horizon!

In spite to its unquestionable rigor, the Bohr theory did not describe chemical bond. In contrast, the "soft" Thomson concept has enlightened the way for the future science in the search for the acceptable picture about the electronic structure of molecules. It should be pointed out that Bohr theory has also additional defect. Suppose that the Bohr calculation of the bond energy could be improved by using, for example, the more accurate starting numerical parameters. The result would be better, but not substantially. It would still be far from the experimental observation. Substantial improvement of this theory that could give more accurate values of energy of chemical bond is not possible, because it is limited by the basic classical-mechanical axiomatics. New paradigm, which must reshape the theory in a way that it would be able to predict physical observations in broader sense, is in sight. The new system should be, by its epistemology, open, what means that the calculations should have the form to be iteratively refined until they will describe experimental observations with satisfactory accuracy.

As it is known from the history of science, especially from the history of physics, this new system is quantum mechanics, which has been for the first time formulated by Werner Heisenberg (1925), and in year latter Erwin Schrödinger. Heisenberg's and Schrödinger's quantum mechanics is in principle a sort of mathematical-metaphysical trick that has not only changed the fundamental world view, but also the general way of thinking. In continuation of this lecture, I will try to demonstrate how almost analogous scientific revolution with similar metaphysical consequences has appeared in chemistry ten years before it has aoppeared in physics. Since these lectures are focused on philosophy of chemistry, I will focus your attention to the paradigmatic change within chemical conceptual systems. Partially, I would try to refer to the common conceptual ideas with physical quantum mechanics.

My represented picture of the science, which is positioned between physical and chemical poles, will be reflected also in this scientific revolution. As the discovery of atoms has triggered the polarization between scientists in their opinion about the physical reality of atoms (physical and chemical atoms in the tenth Lecture), the description of electrons can also have its physical and chemical version.

Heuristic concept of "chemical electrons", which has helped not only in explanation and prediction of geometries (shapes) of molecules but also in description of a series of other properties, especially their chemical behaviour represented within the frame of the *reaction mechanism concept*, has appeared as an extension of the Thomson model. Hundred years ago, 1916, Gilbert Newton Lewis has published in the Journal of the American Chemical Society the paper entitled *The Atom and the Molecule* [1] in which chemical bound is described as a pair of electrons. The basic idea in this Lewis approach is that *electrons do not have identity* (in contrast to Thomson's concept about electrons). Thus, in common electron pair in chemical bond it is not possible to determine the origin of the particular electron. Such an approach, as well as the opening of the problem of identity, is characteristics of the later developed quantum mechanics.

Lewis idea about electron pairs is the direct consequence of the previously described Abegg octet rule. As I have mentioned, Thomson has already suggested that electrons are in atoms divided in two groups, the inner electrons, and the outer electrons which are responsible for the formation of chemical bond. Inspired with the octet rule, Thomson has proposed the hypothesis that the maximal number of outer electrons can be eight. **Walther Ludwig Julius Kossel** (1888–1956) has in *Annalen der Physik* in the same year (1916) published the discussion in which the octet rule (regarded as eight outer, say, *valence electrons*) is explained within the Bohr theory of electrons [8]. Eight Kossel's electrons are orbiting between two atoms which form chemical bond. By Kossel, this orbit is shifted closer to the atom that is more electropositive, in accordance with the Thomson idea about the polarity of molecules. In this way, for instance, eight electrons in the hydrogen chloride molecule, HCl, are in the orbit that is closer to the more positive hydrogen atom, or, in the CaO molecule electrons are orbiting closer to calcium atom. If the molecule consists from two equivalent atoms, as it is the nitrogen molecule, $N_2$, the electrons are orbiting in the plane which is in the middle of the interatomic distance between nitrogen atoms. The complete picture of electronic structures of molecules was based on the idea about the structure of atoms of noble gases, which do not form chemical compounds because they already possess eight outer electrons, and they are in some way "saturated". As the best example for this theory Kossel has used the argon atom.

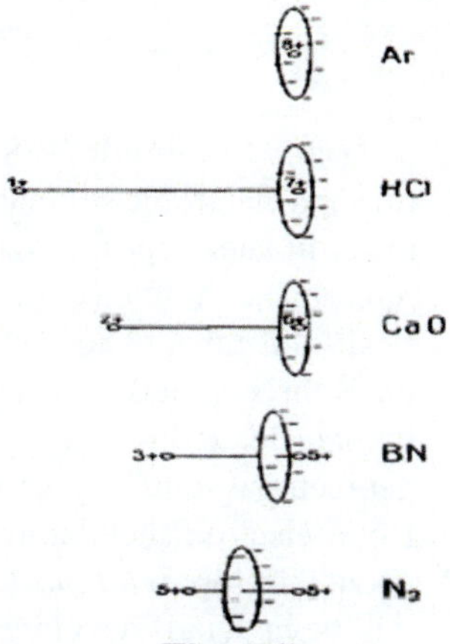

Fig. 8. ¹)

**Facsimile from the work of Walther Ludwig Julius Kossel about the electron structure of molecules. Eight electrons are orbiting in the plane close to the atomic nucleus of more electropositive element [8] (Kossel 1916)**

The lightest noble gas, which would be the best reference, helium, has been neglected because its electronic structure was unknown at that time. However, another important discovery has played the crucial role in the development of Lewis electronic theory.

**Alfred Lauck Parson** (1889–1970) has explained chemical bond as a magnetic phenomenon. In his book from 1915, Parson has proposed that electron behaves as a sort of magnet, because it is known that the movement of electric charge induces the magnetic field [9]. If electrons are "magnetized", it could be expected that atoms can bind to each other because their electrons, which behave as magnets, mutually attract with their magnetic dipoles oriented to the opposite directions. In this way, Parson has resolved the problem of the formation of chemical bond between the identical atoms in spite to the fact that between them do not extst the difference in electric polarity, the problem that origins already from the dualistic concept of Berzelius. Since *magnetons*, as Parson calls electrons-magnets, can in the vicinity of atomic nucleus be distributed in several configurations, the molecules can possess different shapes and symmetries. Evidently inspired with the octet rule, Parson has proposed that the best combination can be obtained with eight magnetons.

**Photo of the model from Parsons book, which demonstrates the optimal space configuration of eight *magnetons* (electrons which behave as magnets)**

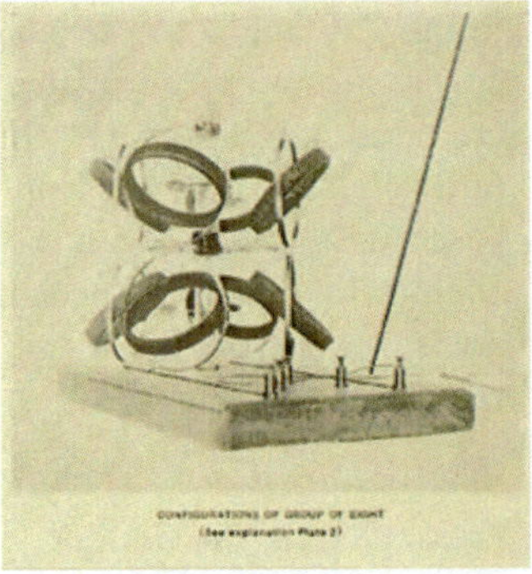

**Irving Langmuir** (1881–1957) has also studied the relationship between the octet rule and electronic structure, but with the special focus on the *symmetry*, the category that will later appear as one of basic concepts in quantum mechanics. Langmuir has represented eight electrons symmetrically distributed on the vertices of the geometrical solid cube. In his picture, electrons are in atoms divided in shells (similarly as in the Bohr's model), and these shells *differ between each other by symmetry*: each shell is characterized with its particular distribution of electrons that their position form the geometrical solid with various number of planes, or axes of symmetry [10]. Since, by Bohr, electron shells differ in energy of their electrons, Langmuir propose the idea about the close *relationship between energy and symmetry*, the theoretical model that will be in details developed within quantum mechanics.

Such model about eight electrons in vertices of cube has been accepted by Lewis, who has refined this concept by pairing electrons in four electron pairs. Lewis emphasis on the electron pair instead of the electron octet was a consequence of the investigation of the electron structure of helium, which, in contrast to other noble gases, possesses only two electrons.

By Lewis, the electron pair which forms chemical bond is necessary condition for the accomplishing the electronic octet. Taking into account Parson-Langmuir ideas about cubic symmetry, Lewis has proposed representation of the electronic structure of molecules as interlinking of cubes in which electrons are situated in vertices. Atoms of chemical elements which have less than eight outer (valence) electrons are modelled with cubes with so many empty vertices how many electrons they miss to fulfil the octet, as it is clearly shown in original Lewis drafts about the electron structures of some atoms.

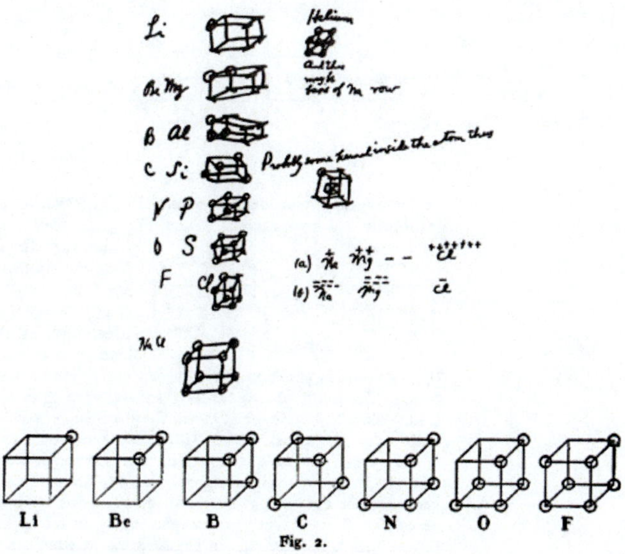


**Facsimile of Lewis representations of distribution of electrons in vertices of cube for different atoms [1] (Lewis 1916)**

The upper figure demonstrates that, for instance, lithium atom (Li), which possesses only one electron in outer shell, is represented as a cube with only one vertex occupied with electron. Following the same rule, beryllium (Be) has two occupied vertices, boron (B) three, etc., till fluorine with seven electrons on vertices.

Within this model, Lewis shows structures of molecules as interlinking the cubes (the representations of atoms) either with common vertex, common edge, or, common plane, what depends whether the *chemical bond* is *single*, *double*, or *triple*. In the following original paragraph, Lewis describes the oxygen molecule.

**Facsimile of Lewis representations of interlinking of atoms modelled as cubes with electrons in vertices**

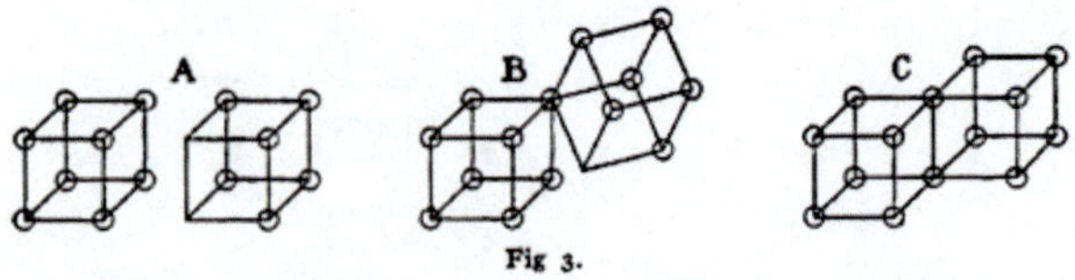


Fig 3.

We may next consider the *double bond* in which four electrons are held conjointly by two atoms. Thus Fig. 4, *A*, may represent the typical structure of the molecule of oxygen. A characteristic feature of the double bond is its tendency to "break." When this happens in a symmetrical way, as it will, except in a highly polar environment, it leaves the two atoms concerned in the *odd* state, each with an unpaired electron in the shell. In so far as a substance with a double bond assumes this other tautomeric form, it will show all the properties of the substances with odd molecules. Thus Fig. 4, *B*, represents this tautomeric form of the oxygen molecule; the equilibrium between forms *A* and *B* is entirely analogous to the equilibrium between $N_2O_4$ and $NO_2$. At low temperatures almost every known case of combination with oxygen gives first a peroxide. This shows that oxygen exists to an appreciable degree in a form which approximates to the form *B*, in which it can add directly to

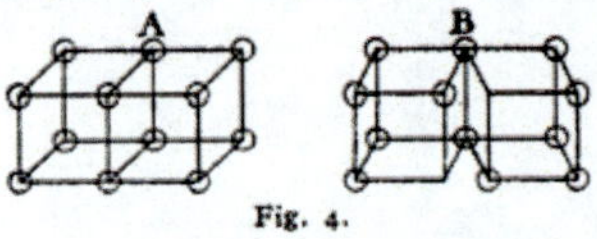


Fig. 4.

After Lewis, the chemical formulas, which were developed starting with the mass equivalents of elements, have gained another meaning. Now, *bonding lines* between the symbols of elements became the *symbols for the electron pairs*.

**H ·     · H     H : H**

is the same as the formula **H-H**

This symbolism is also shown for the molecules of methane and etene:

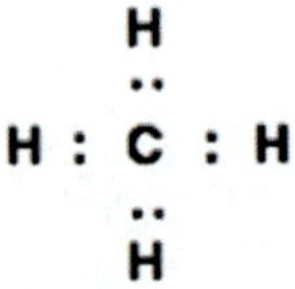

Since in some molecules, as for example in the molecule of ammonia, two of electrons are not included in covalent bonds, Lewis has introduced the category of *nonbonding electrons*, or the *lone electron pairs*. In the following figure, the electronic structure of ammonia is represented with the lone electron pair above the symbol of nitrogen atom.

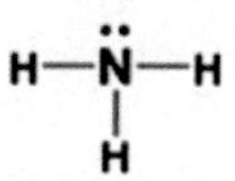

Just this concept of electron pairs, especially of lone pairs, will have fundamental importance for the successive development of the theory of reaction mechanisms. Starting with its own idea about electrons as entities without their identity, Lewis understand that electrons could from that reason be formally redistributed in the molecule by changing their role of bonding, or nonbonding electron pairs. Evidently, electrons are not only without their identity (i.e., the secondary identity, in accordance with my previous lectures) but also without the defined position in molecule—they are *delocalized*. This concept of delocalization can be clearly recognized from the Lewis original papers. The real, physical electronic structure of molecule is modelled as a set of representations with different electron configurations, as it is shown for the molecules of oxygen and etene:

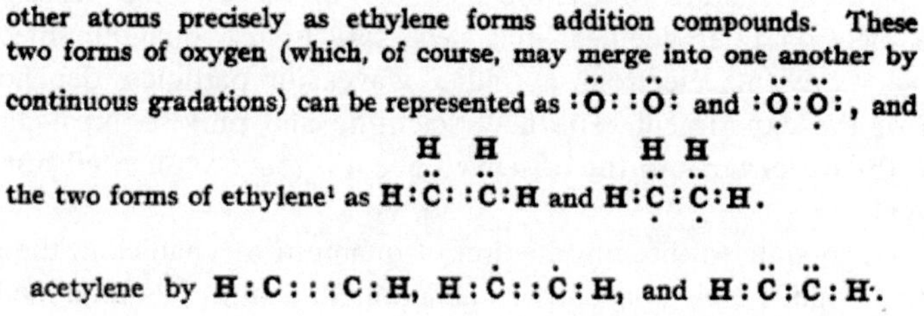

**other atoms precisely as ethylene forms addition compounds. These two forms of oxygen (which, of course, may merge into one another by continuous gradations) can be represented as :Ö::Ö: and :Ö:Ö:, and the two forms of ethylene[1] as H:C::C:H and H:C:C:H.**

**acetylene by H:C:::C:H, H:Ċ::Ċ:H, and H:C̈:C̈:H.**

This representation of a molecule by using more than one formula is a far-reaching vision: it is metaphysical background for the concept about *superposition of configurations*, the concept that has later been developed as one of the crucial statements in quantum mechanics.

I cannot understand why this Lewis concept has been neglected, and why it has never been appeared in chemical textbooks! As the consequence, in most of the textbooks the Lewis electronic structures of some molecules were erroneously represented. For instance, the oxygen molecule is in textbooks represented with only one structural formula in which every oxygen atom has two lone electron pairs, contrary to the original Lewis model shown on the top of the upper facsimile. Such wrong interpretation of Lewis theory has caused a series of misunderstandings in scientific discussions, especially in the discussions about the nonclassical reactive intermediates [11].

His theory about electron pairs Lewis has extended also in explanation of the stereochemistry of organic carbon compounds. In this approach, the electron pairs are redistributed in the opposite edges of the cube by forming tetrahedron.

**Figure from Lewis work about the tetrahedral distributions of electron pairs in the molecule of methane**

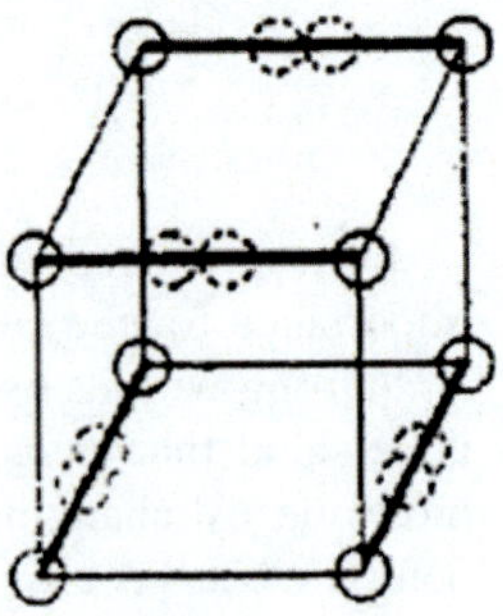
Fig. 5.

Although Lewis concept, as well as the later developed model of resonance, form the basis for unequivocal use of the structures for the classification of organic compounds and understanding chemical changes, it was still insufficient for the explanation of the nature of chemical bond. For the detailed description of the nature of electron in molecules, it is necessary to use quantum mechanics, the theory that represents electrons as either waves, or particles, depending on the selection of the experiment. The new scientific and philosophical paradigm has appeared: **the observer and the observed are not more separated but coupled to each other!**

However, in spite to the introduction of quantum mechanics in the explanation of chemical bond, Lewis concept was abandoned. Still, it has remained one of the basic models in chemistry. **Christopher Kelk Ingold** (1893–1970) and **Robert Robinson** (1886–1975) have created the idea that, perhaps, not only the molecular structures, but also chemical reactions, which are in principle merely the changes of these structures, could be represented by using Lewis electron pairs. If chemical bonds, which define constitution and configuration of molecule are electron pairs, then chemical reaction regarded as breaking and formation of chemical bond can be represented just by shifting electron pairs from one to another position in the molecule. Ingold and Robinson have named this model the *theory of reaction mechanisms*.

The historical way to this theory was not the straight line. Rather, the concept of reaction mechanism was created through a series of phases. The mechanistic theory has its origin already at the time of the first, "nonelectronic", structural concept. Close to the end of the 19th century, **Arthur Lapworth** (1872–1941) tried for the first time to describe chemical reaction starting from the structural theory. As the model for his theory, Lapworth has studied the reaction of addition of cyanide ion to carbonyl group in benzaldehyde. The basic experimental method that Lapworth has used was chemical kinetics, i.e. the investigation of the rate of chemical reaction. By patient measuring the rates of addition of cyanide ion to the carbonyl group, Lapworth has concluded that this reaction occurs in two steps. The first step yields the short-living unstable ion with the negative charge on the oxygen atom,

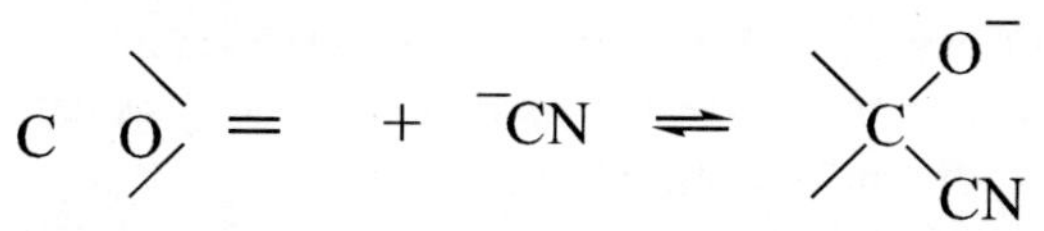

and in the second step the positive hydrogen ion (proton) binds to oxygen forming the final product—cyanohydrin. In this explanation, two of new notions were introduced: the *reaction mechanism*, and the *reactive intermediate*, respectively. Such concept of reaction mechanism is the core of the concept in which chemical reaction, i.e. the transformation of molecular constitution (and configuration), can be divided in the sequence of simple elementary reaction steps. Every step represents a minimal change in configuration on the pathway from the reactants to the products. Some of molecules that could appear in particular step can have specific structures that can survive only very short period of time—*the reactive intermediates*. In the simple example examined by Lapworth, the reaction occurs through only one intermediate (with the negative charge on oxygen). As I have mentioned, the second step yields the final product:

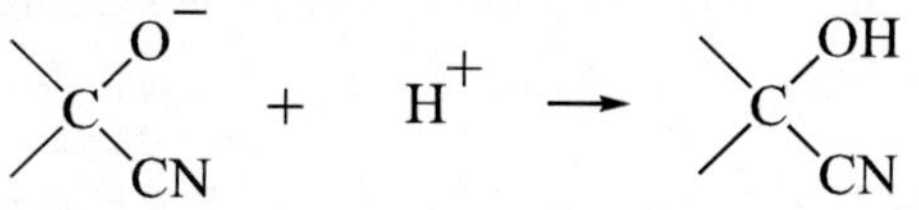

For the explanation of this mechanism, Lapworth has proposed the *ad hoc* hypothesis by which all the atoms in molecule possess a sort of "*latent polarity*" that is activated only in the moment of collision of the reacting molecules. In the represented example oxygen has negative latent polarity, which attracts proton, $H^+$. This hypothesis has been supported by the isolation of the reactive intermediate in the form of sodium salt:

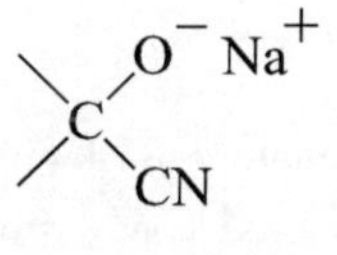

Latent polarities of atoms can be predicted in almost all organic compounds known at this time. Positive or negative charges were determined by using the principles developed earlier within the dualistic theory and electrochemical experiments. Distribution of latent polarities is always alternating—in the neighbourhood are always oposite charges, as it is shown in the example:

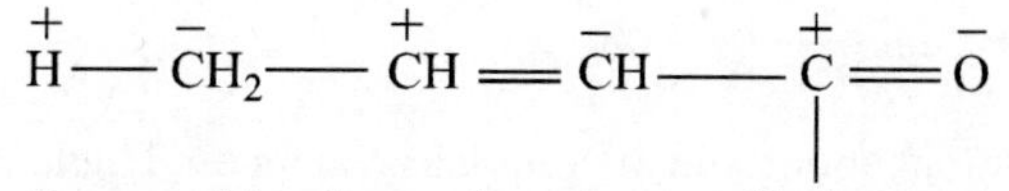

Because of that, Lapworth's theory has been called the *theory of alternate polarity*. This theory was successful because it has explained a series of reaction mechanisms. For instance, **Johannes Thiele** (1865–1927) has used the method of alternate latent polarity for explanation of the mechanism of addition of methyl-iodide to the unsaturated ester:

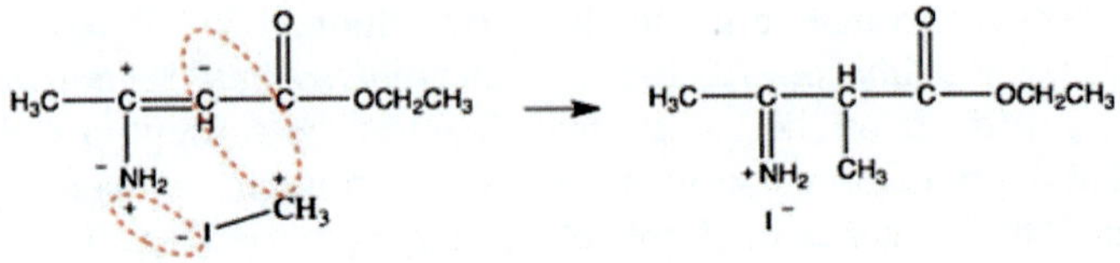

In the moment of collision, the molecules are polarized in alternate charges in such a way, that the iodine atom becomes negative, and the methyl group becomes positive. Tis combination of charges yields the final product.

However, Thiele's contribution to the theory of reaction mechanism is more important because of his investigation of the conjugated polyenes (hydrocarbons with the alternate single and double bonds between the carbon atoms). The addition of hydrogen on such conjugated polyene can be represented in a following way:

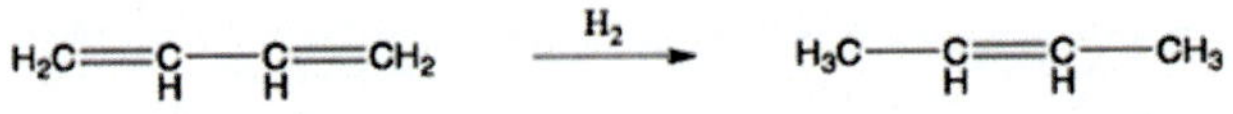

Thiele has recognized that after the reaction, thus, in the product, the double bond is shifted to another position, relative to its position in the starting polyene. In the case of the longer molecule this phenomenon is represented as:

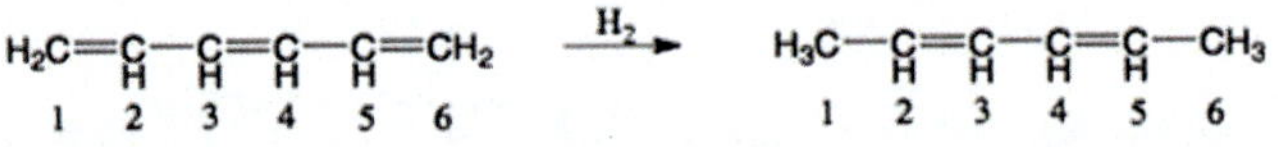

Before the reaction (in the reactant molecule) the double bonds were situated between the atoms $C_1C_2$, $C_3C_4$, mad $C_5C_6$, but after the reaction (in the *product* molecule) the double bonds appear between atoms $C_2C_3$ and $C_4C_5$. From this observation, Thiele has extended the Lapworth concept with two important improvements. First, in the represented reaction there is a shift of double bonds, what means that **these double bonds are "movable"**. Second, Thiele has introduced the *ad hoc* hypotheses that conjugated polyenes possess a sort of "*incomplete valences*".

This speculation about the incomplete valences Thiele has also applied for explanation of the Kekulé's paradox about the behaviour of the benzene molecule by proposing that, besides the three "normal" double bonds, the molecule possesses also three incomplete valences:

Since in the Lewis theory, the chemical bond is described with electron pairs between the two interconnected atoms, the mobility of double bonds should be replaced with mobilities of electrons. In this picture, electron pairs can move within the molecule, but they can also be exchanged between two reacting molecules. Lewis electronic theory affords in this way the possibility of its extension, from the explanation of molecular structure to the explanation of changes of this structure during chemical reaction. Mobility of electron pairs has occurred to be a starting point of the concept of reaction mechanisms, which was later developed by Robinson and Ingold, and which became one of the pillars of modern chemistry. Robinson has unified Thiele's concept about shifts of double bonds with the Lewis's electron pairs theory in the hypothesis about the "ease relocation of electrons in unsaturated systems". As the constitution of molecule is in chemical formulas represented with bonding lines, which are symbols for electron pairs, chemical reaction is in analogous way represented with arrows as labels for shifts of electron pairs [12]. In the literature, these are known as the "Robinson arrows".

In addition, Robinson has used this concept for the explanation of the structure of benzene. If Thiele's *incomplete valences* are replaced with electron pairs, then the double and the single bonds in the benzene molecule cannot be distinguished: the *electrons are delocalized* over the whole six-membered ring. Robinson represents benzene in two ways. The first picture is the six-membered ring with the circle in the middle, which represents six delocalized electrons.

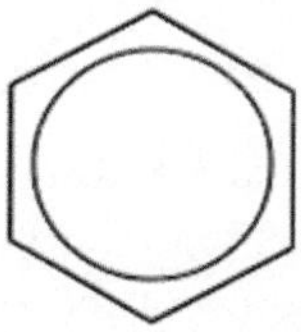

The second description was even more revolutionary. Robinson, understanding that the description of benzene molecule, because of its mobile electrons, is not possible with only one "classical" Lewis formula in which the positions of single and double bonds are determined, has proposed that benzene must be represented minimally with two formulas. He has called them *mesomeric structures*, but later this term is replaced with *resonance structures*.

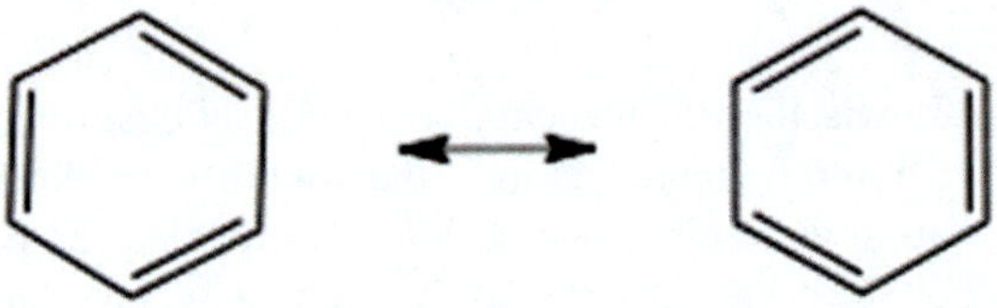

If the molecule is better represented with more structures than with only one, its "real structure", or better to say its physical state, is a sort of "superposition" of larger number of constitutions represented by Lewis structures. It is not quite clear, whether Robinson was aware that, by proposing this concept he has obviated one of the fundamental statements of quantum mechanics about the superposition of all possible states.

Although physics and chemistry were developed from quite different origins, as I have discussed in previous lectures, both the disciplines have approached to each other by proposing the same metaphysical representations of the universe.

It must be pointed out that Lewis theory has been not extended only in the form of the new theory of reaction mechanisms, but also with new investigations of the "exotic" thermodynamically unstable molecules, which appear as reactive intermediates in chemical reactions. Many of these molecules cannot be represented with the Lewis structures based on the electron pairs between the bound atoms. These structures are known as *nonclassical*. For instance, the molecule of *2-norbornyl cation* is described with the tricentric-twoelectron bond. That means that three carbon atoms (not only two as in Lewis structures) are bound to each other with two electrons:

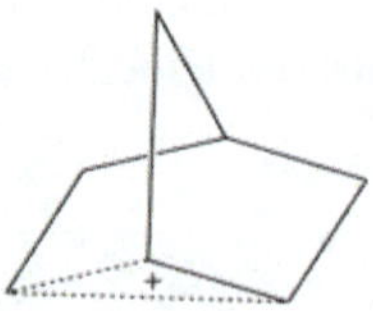

Christopher Ingold and **Edward David Hughes** (1906–1963) have extended and finalized the modern concept of reaction mechanisms based on the mobility of electron pairs with the classification of reaction mechanisms. This systematization does not include only the *classes of chemical compounds* but also the *classes of chemical* reactions [13]. In this way, chemical systematization became "two-dimensional"; in one direction, it is classification of substances that is based on the Lewis-type structures of their molecules with fixed electron pairs between atoms, and in perpendicular direction it is the classification of chemical reactions that is based on the movement of these electron pairs.

These two classifications are new challenge not only for further research, but also for the teaching methods in chemistry. Appearance of two "perpendicular" systematics in science denotes the difficulty for education. In modern teaching of chemistry these methodical paradoxes are not solved, and they cannot be removed.

Till the early sixties, the organic chemistry textbooks were based on the classification of the types of substances. That means that in these textbooks we can find chapters about, for example, alcohols, ethers, aldehydes, amides, etc. Thus, in these textbooks any information about the important properties of particular organic compound can easily be found. However, if you wish to propose the series of reactions by which the new chemical compound can be synthesised, these reactions are difficult to find in this textbook, because they are dispersed through various chapters.

New approach to education in organic chemistry has appeared in 1959 with the textbook written by **George S. Hammond** (1921–2005) and **Donald J. Cram** (1919.–2001.). Later, in the new edition, 1970, the third author **James B. Hendrickson** joined this book [14] (Hendrickson, Cram, Hammond, 1970). In this textbook organic chemistry is classified by types of chemical reactions in accordance with the works of Ingold and Hughes, and not by types of chemical compounds. There are chapters entitled as nucleophilic substitution, additions, eliminations, molecular rearrangements, etc. The advance and the scope of this textbook is that students are well educated in chemical synthesis, in designing of novel molecules, new substances, because it develops the skill for combining typical chemical reactions. However, the information about particular organic compounds and their properties is in this textbook almost unreachable because it is scattered all over the entire contents.

The conclusion is interesting. If we decide to receive one kind of information (for instance about particular substance), another kind of information (how to combine chemical reactions to synthesize the target compound) is unavailable. This is a sort of the principle of exclusiveness, which could also be applied out of chemistry, even out of science, and which tells us that *the picture of the world, or perhaps the picture of the Universe as itself, depends on the decision of the observer!*

There are many analogies for this principle. The subject decides what to observe with the awareness that by this observation some other information vanishes. Practical example is represented in the following photos. By photographing the car in movement with very short exposition on camera, say 1/1000 seconds, the photo provides information about the exact position of the car (left photo), but from this photograph no information about the speed of this car is available. On the other hand, if we take the photo with the long exposition, say two seconds, we get the picture (right photo) from which we cannot get information about the position of the car. However, from the smear on the photo we can (knowing the exposition) calculate the speed of this car. Thus, if we decide to determine the position of the car, the information about its speed vanishes, and vice versa, if we decide to measure the speed, the information about position vanishes. *Quod erat demonstrandum*!

**Photos that could serve as a metaphor of the idea about the uncertainty principle. For the car on the left photo we know position, but the information about its speed is unavailable. From the length of the smear on the right photo we can calculate the speed, but we don't know the position of this car**

The dependence of the response of nature on the observer's choice, i.e. on the coupling of subject-object, could clearly be recognized in the choice of chemist to observe molecules either through Lewis structures, or through quantum mechanics. In the methane molecule, $CH_4$, there are four equal chemical bonds of hydrogen atoms with carbon. These four Lewis electron pairs is a good representation because it informs us about the tetrahedral geometry of the molecule. However, quantum mechanics shows that four electron pairs which have the same energy do not exist in the methane molecule. Instead, six electrons have higher energy, and two electrons have lower energy. If we decide to use Lewis model, we can obtain the information about the shape of the molecule, and the information about the fact that all the carbon-hydrogen bonds are equal, but the information about electron energy is missing. Conversely, quantum mechanics in the form of a molecular orbital model, interprets well the electron energy, but discourages us in interpretation of geometry starting from the concept about the four equal electron pairs in carbon-hydrogen bonds.

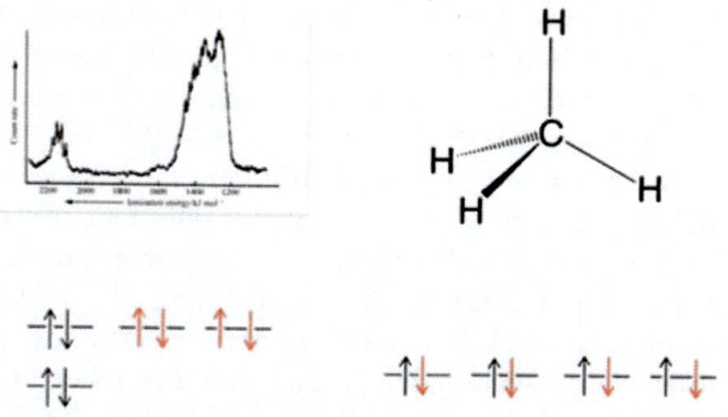


**Representation of electron distribution on the left picture provides information about the electron energies (what is also supported with the photoelectron spectrum on the top of the figure), the information about the geometry is vanishing. Representation of the electron distribution on the right figure is wrong, but it helps to understand the tetrahedral geometry of the methane molecule**

Alchemical textual matter has been written in the secret language and idioms. Two contents were present in the same text, the operative and the speculative component. If we decide to follow the operative interpretation, we could probably be able to repeat the alchemical experiment, but in this moment, the hermetic contents will vanish. Conversely, the hermetic, speculative interpretation excludes any information about the practical manipulation with substance. Aristotle with his concept of deprivation, the possibility of change (στερεσις) is already in this linc, because he supposes that by any change, for instance by melting in oven, all the spectrum of possibilities vanishes except the particular transformation selected by the observer—the alchemist.

From the very beginning, from Heisenberg uncertainty principle and Schrödinger equation, quantum mechanics intended to confirm its principles by explanation of chemical phenomena, especially the question about chemical bond. The first calculation of the simplest molecule by using Schrödinger equation, which has appeared in 1927 when Heitler and London were calculated the structure and energy of $H_2^+$, has been developed in the method known as *valence bond method* (VB) [5]. Since the mathematical procedure within the VB method is too complex for calculations of larger molecules, **Friedrich Hermann Hund** (1896–1997), **Robert Mulliken** (1896–1986), **John C. Slater** (1900–1976), **John Lennard-Jones** (1894–1954) and **Erich Hückel** (1896–1980) were developed another method called the method of *molecular orbitals* (MO) . Central, and in some sense unified theory of chemical bond has published **Linus Pauling** (1901–1994) in his work *The Nature of the Chemical Bond and the Structure of Molecules and Crystals* (1939).

The problem which both the methods should overcome is the infinitesimal complexity of wave function as the representation of superposition of all the possible states. However, such wave-function could be approximated with an algebraic series of simpler functions. Since even such calculus is very complicated, it could be further simplified by introducing some parameters used from the *experience*, i.e. some measured numerical values. Thus, the empirical experience turns out to be a component of calculation—science becomes *semiempirical*. Finally, calculations lose its universality and become more useful if they are accommodated to particular classes of problems. In this way, the universal questions about the structure of matter were (at least partially) lost from the main stream in science. Rather, the investigations are, with their practically oriented methods, predominantly oriented to the concrete objects. In philosophy of science appears new directions such as *empirical constructivism* within which **Bas van Fraassen** [15–17] argues that science is acceptable if it should explain particular phenomena, not the universal principles. Paradigmatic change that has appeared in Copenhagen school, which has introduced the idea about the "dual nature" of matter, i.e., the inseparable coupling between the observer and observed, has been transformed to the empirical constructivism, which, in unskilled hands can become allegation for superficiality. In this way, cosmological questions about the macro- and microcosm are in science and education replaced by actual everyday life. Is the new science merely a discipline subordinated to the need for simple things, something "higher" because of something "lower"? Though, if we look on the science as on the organized scepticism rather than to a tool for getting a sort of final solutions, there is a hope that these tendencies could led to new paradigm.

## 12.1 Conclusions

The Lewis' formulation of chemical bond as an electron pair is the basis for modern practical structural theory and in the same time the explanation of the meaning of chemical formula. However, the extension of this concept to the study of chemical reaction has two important consequences. First is the development of the theory of reaction mechanisms, and the second is the introduction in science the general concept about the superposition of representations. This new paradigm has appeared in parallel with the formulation of quantum mechanics in physics.

It is also demonstrated how the uncertainty principle that appeared originally in quantum mechanics could be used as general epistemic model in chemistry, chemical education, and in practical life.

## References

1. Lewis GN (1916) The atom and the molecule. J Am Chem Soc 38:762
2. Nernst W (1913) Richard Abbeg, Berichte d. D. Chem.Geselschaft, Jahrg. 46:40
3. Bohr N (1913) Philos Mag 26:857
4. Heitler WH, London FW (1927) Wechselwirkung neutraler Atome und homöopolare Bindung nach der Quantenmechanik. Zeitschrift für Physik 44:455

5. Duhem P (1991) The aim and structure of physical theory. Princeton University Press, Princeton, NJ
6. Gijsbers V (2014) Unification as a measure of natural classification. THEORIA 79:71–82
7. Thomson JJ (1914) The forces between atoms and chemical affinity. Philos Mag S.6. 27:757
8. Kossel W (1916) Über Molekülbildung als Frage des Atombaus. Ann. der Physik 49:229
9. Parson AL (1915) A magneton theory of the structure of the atom. Smithsonian, Washington
10. Langmuir I (1919) The arrangement of electrons in atoms and molecules. J Am Chem Soc 41:868
11. Brown HCh, Schleyer PvR (1977) The nonclassical ion problem. Plenum Press
12. Robinson R (1932) Two lectures on an "Outline of an electrochemical (Electronic) theory of the course of organic reactions", The Institute of Chemistry of Great Britain and Ireland, London; J Soc Dyers Colourists, Jubilee Issue, 1934, 65
13. Ingold CK (1953) Structure and mechanism in organic chemistry. Cornell, University Press, Ithaca
14. Hendrickson JB, Cram DJ, Hammond GS (1970) Organic chemistry, International edn. McGrow-Hill Kogakusha, Tokyo
15. van Fraassen B (1980) The Scientific Image. Oxford University Press, Oxford
16. van Fraassen B (2006) Structure: Its Shadow and Substance. Br J Philos 57:275
17. van Fraassen B (2010) Scientific representation. Oxford University Press, Oxford

# Teleology

# 13

*Definition of design = Everyone designs who devise courses of action aimed at changing existing situations into preferred ones.*

Herbert A. Simon

**Abstract**

This chapter is a reflection of the first lecture, and it deals with the problem of chemistry in designing of new substances, which have its scope in the appearances on the levels of complexity higher than chemical level. The problem of chemistry as basic science, and chemistry as auxiliary discipline is discussed.

**Keywords**

Structure-function relationship · Teleology in science · Curiosity · Academic freedom · Antropic principle

Ladies, and Gentlemen,

**Michael Polany** (1891–1976) is very well known as a chemist who gave important contribution to the development of the transition state theory of chemical reaction. Its BEP principle (Bell-Evans-Polany) helps chemists to estimate some of the important properties of the transition states, and to predict chemical reactivity. However, his intellectual activity was much broader. Polany has been engaged in discussions about the philosophical and sociological position of science, academic freedom, and university education [1].

I wish to represent how the work of Polany could give a new light to the position of chemistry on the scale of complexity levels, [2] that I have discussed in the first



H. Vančik, *Philosophy of Chemistry*, Integrated Science 2,
https://doi.org/10.1007/978-3-030-69224-7_13

lecture. Let me start with analogy. Only humans can construct machines, the artefacts consisting of interrelated parts. But the crucial property of machines is their human purpose. Design of machines is based on the *restrictions* which we as humans give to the nature, say to natural laws, the *boundary conditions*. These boundary conditions are imposed to laws of physics to get such a functional construction, which will have a desired scope. All our designs have their purpose in contributing to particular practical requirement of our life. Now, we can switch our sight from analysing of the nature to the designing of the nature. From our position as humans, which is on the top level on the complexity scale, everything below this level must have its purpose for us. This point of view that everything in the nature must have its purpose is known already from Aristotle who has called it *teleology* (τέλος, purpose, scope).

The question is whether teleological principle is only a matter of humanity, or it is a principle of more general significance. If we pass from machines to living beings, then there are not mechanical parts and the overall construction, but the organs that have their function, and chemical background that supports and enables this function. More simply, let us take enzymes and their *biological function*. Enzymes were formed as a consequence of the evolutionary pressure, i.e. the *boundary conditions* in such a way that they can play particular role in living organism.

Now we have the interplay between two complexity levels, biological one that requires function for its existence, and the chemical level, which enables this function if it is suppressed from the upper (biological) level. Does it mean that lower complexity levels are in teleological sense determined by particular purposes on the higher complexity level? This is not the naïve question because the main stream in the contemporary research in chemistry is focused on the *structure-function relationship*. Chemists have to synthesize, to design new substances whose structure would enable function, for instance in medical treatment. Many doctoral theses in organic chemisrty are entitled such as "structure function relationship between….". However, the chemists are not aware that the "function" is not a matter of chemistry, but rather it belongs to the discipline on the higher level of complexity. For instance, the function of haemoglobin is to transfer oxygen to cells, but this is not the chemical question, rather it is question of biology. Chemical question is not this function but the molecular structure of haemoglobin, the reaction of this molecule with oxygen or with other molecules. Here we have functional and non-functional explanation [3]. In this way, chemistry is on borderline with biology. The same ontological problem appears with designing of new technological materials, which could have their either chemical or functional explanation.

We as humans adjust boundary conditions to chemical laws to accommodate the nature for us. Is it a variant of the *antropic principle*, which states that the evolution of universe inevitably led to the appearance of intelligent beings?

As I have discussed in the first lecture, the relation between complexity levels are determined by causality. Structures on the lower level of complexity are "causes" for the behaviour of the entities on the higher level, not oppositely. Atomic structure is a cause that molecules have corresponding properties, but molecular emergent properties cannot change atomic structure. But now, in this lecture, we have argued also that the higher level has effect on the lower level by setting up the boundary conditions that enable the purpose, the scope of the higher-level entity. We as humans appear on the top complexity level, and all we are doing has particular goal for which we set up the boundary conditions to science as such.

In this frame, chemistry opens the philosophical, or, better the sociological question of general importance: what is the relation between the fundamental and the applied research? If the applied research is determined in the top-down way, i.e. from the requirements of human existence, then the main stream in today's science is justified. What is the basic science in this context: the science triggered merely by *curiosity*? But on the other hand, curiosity, as well as aesthetical requirements, belong to the most sophisticated characteristics of man. Curiosity results with explanation, understanding, and new visions on the nature and society. Science has its aesthetical component, the beauty of theories, the beauty of representations or mathematical acrobatics. Three-dimensional representations of structures of molecules and crystals are the best example for this.

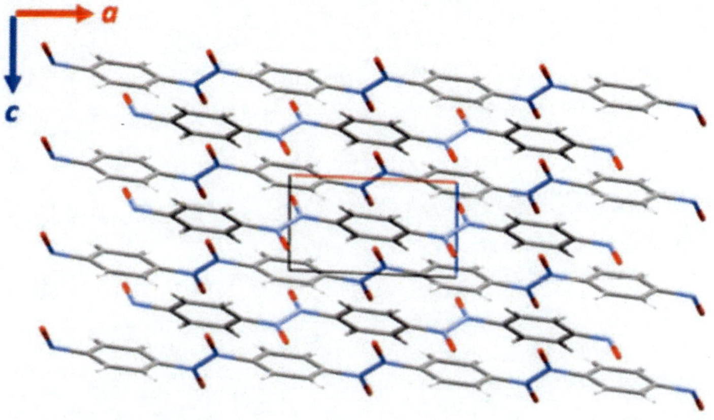


## 13.1 Conclusions

In contrast to most of other natural sciences, chemistry is not only the analytical discipline, but also synthetical one. Chemists prepare the objects for their investigation, thus, chemistry as a science has a teleological component. The modern main-stream chemistry is focused on the structure-function relationship. However, I argue that while structure is a chemical category, its function (for instance biological) is not. Chemical structure and biological function have quite different explanations because they belong to different complexity levels of the nature.

## References

1. Hartl P (2012) Synthesis Philosophica 54:307–321
2. Polany M (1968) Science 160:1308–1312
3. Woody A, Glymor C (2000) in: Of Minds and Molecules, N. Bhushan, S. Rosenfeld, Eds. Oxford University Press, p. 17 – 34.

# Author Index




H. Vančik, *Philosophy of Chemistry*, Integrated Science 2,
https://doi.org/10.1007/978-3-030-69224-7

# Subject Index




H. Vančik, *Philosophy of Chemistry*, Integrated Science 2,
https://doi.org/10.1007/978-3-030-69224-7